Geo-epistemology

Hispanic Studies: Culture and Ideas

Volume 23

Edited by
Claudio Canaparo

PETER LANG
Oxford · Bern · Berlin · Bruxelles · Frankfurt am Main · New York · Wien

Claudio Canaparo

Geo-epistemology

Latin America and the Location of Knowledge

PETER LANG
Oxford · Bern · Berlin · Bruxelles · Frankfurt am Main · New York · Wien

Bibliographic information published by Die Deutsche Nationalbibliothek.
Die Deutsche Nationalbibliothek lists this publication in the Deutsche Nationalbibliografie; detailed bibliographic data is available on the Internet at http://dnb.d-nb.de.

A catalogue record for this book is available from The British Library.

Library of Congress Cataloging-in-Publication Data:

Canaparo, Claudio.
Geo-epistemology : Latin America and the location of knowledge / Claudio Canaparo.
p. cm. -- (Hispanic studies, ISSN 1661-4720 ; 23)
Includes bibliographical references and index.
ISBN 978-3-03911-573-0 (alk. paper)
1. Social epistemology--Latin America. 2. Knowledge, Theory of. 3. Geography--Philosophy. 4. Geography--Latin America. 5. Knowledge, Sociology of. I. Title.
BD175.C36 2009
121--dc22
2009023988

Graphic design: Florian Ziche.
Cover illustration: 'Localidad' © Edmundo Benján, 1970.

ISSN 1661-4720
ISBN 978-3-03911-573-0

Hochfeldstrasse 32, CH-3012 Bern, Switzerland
info@peterlang.com, www.peterlang.com, www.peterlang.net

To Dara, Ethel, and Libi.
For the world we will leave behind.

Cuando miré, ya suicidado,
el paso de las cosas,
escéptico, como olvidado,
encontré atónito rugosas
manchas, objetos, seres,
pero no un pasado.
Y supe que ni enceres
ni vida de olvidado
dejarían el descuido
de visitarme por nada,
por debilidad de ido,
por ausencia ordenada,
por, pores.

Edmundo Benaján
'Pores' from *Manubrios y barrocos* (1984)

Contents

Acknowledgements

I wish to thank Julie Crocker and David Gorman for the painful task of revising the manuscript of this book. I always find their patience, knowledge and efficiency enlightening. All the good things about the style of this book are due to them, and the mistakes still mine.

Regenia Gagnier read a draft of Chapter 3 and I am grateful for her comments about it. I also would like to thank Sergio Plata, Israel San Martin, Kim Schulte, Eduardo Ortiz, Juan Carlos Bayo-Julve, Enrique Spoljaric, Julio Ortega and Sergio Cornejo for their accounts and comments about exile, postcolonial theories, travelling and academia. Also I am in debt to Alejandro Kaufman for his always stimulating challenging comments and with William Rowe for his observations and constant support.

EOLSS Publications from Oxford published an early version of Chapter 3 and I wish to thank them for their permission to include an extended version of it in this book.

Because of the nature and contents of this work, it was reasonable not to expect any support in the academic environment. On the contrary, it was predictable to have some political opposition and few bureaucratic chicanes, as indeed has happened in many ways during the two years I have been writing this work. And, no less important, this situation made me think more of non-academic domains and readers. The result is that I made an attempt, right from the start, to do a book for an educated non-European reader with a basic interest in the idea of knowledge, in colonialism or in the notion of what we call Latin America.

Finally I would like to thanks to Carlos Scharf for his hospitality, advice and libertarian sprit that I enjoyed in many occasions several years ago when the idea about this book first appeared. He was part of a group of anarchist militants in Latin America that have understood the catastrophic cultural consequences of some ideas about colonialism and engagement in political terms.

Brussels, June 6, 2008

Prologue. Spatial Thinking and Foreignness

> J'ai montré comment le même objet placé dans quatre milieux différents prend quatre significations différentes et change chaque fois profondément de caractères. [...] Ce phénomène ne peut recevoir qu'une explication, à savoir que tous les caractères des objets ne sont au fond rien d'autre que des caractères perceptifs qui leur sont conférés par le sujet avec lequel ils ont un rapport.
>
> Jacob von Uexküll (1965: 166)

This is not a book about Latin America. It is a book that implements and assumes a perspective of language and knowledge that should be considered as grounded in the idea of Latin America as homeland. Latin America is not only a subject or an academic argument, it is a perspective on organizing objects, naming an environment, establishing subjectivities, constructing knowledge, producing narratives. And because we will focus on this latter aspect we will refer to this approach as *epistemological*. However, epistemology cannot exist in abstraction, despite the Western/European academic tradition that argues the contrary. And that is also why we use the concept of 'Latin America', to anchor the arguments in a particular location, but not in order to celebrate some folkloric principles.

This book is also a navigation device as it tries to move around all the places that for different authors and theories have been indicated as 'Latin America'. And exactly in this sense 'Latin America' is itself defined as a navigational phenomenon because it is increasingly defined in relation to movement, travelling and 'foreign cultures'.

What is 'familiar' and what is 'non-familiar'? How do we relate one to the other? Why does knowledge presuppose for us this basic distinction between what we consider as 'ours' and what is not part of this domain of property and affiliations? How do we determine the working principle of this *familiarity*? These questions are at the *core* of this book. If we focus on the concept of 'Latin America' and its development, for instance, should we only use sources geographically located 'in Latin America'? What is the meaning of this location?

Should we also consider as 'foreign' what has not been specifically inscribed as 'Latin America'? Is it possible to root the concept of 'Latin America' in different cultural traditions? There is obviously no single answer but a set of perspectives and conditions, some of which we are going to explore in this book.

Of course, writing about a *longue durée* of cultural and conceptual periods requires a level of generality in the arguments that will not satisfy the taste of those readers who expect a detailed and systematic conceptual and chronological perspective. We have done that in the past in relation, for example, to the River Plate area (see Canaparo, 2005). The sentences that refer to extensive periods of time, different cultures and environments and to very complex historiographical narrations should be understood as a necessary *proviso* to delivering the arguments, and not as the final view about the subject.

The concept of *geo-epistemology* that we postulate here aims to address two main issues: (i) one is the meaning of the *location of knowledge*, understood as concept, (ii) the other is the 'épistémo-anthropologique' (Apel, 2005: 21) condition of knowledge, understood as set of concepts and/or as the evolution of concepts. Both issues are obviously related and point to the same conclusion: there is no unproblematic and objective notion of knowledge. Why the idea and/or notion of 'Latin America' should have a particular relation to this concept of geo-epistemology is something that this book attempts to establish.

Finally this book is about *dislocation*, that is, about a regular experience of displacement and being out of place. Many communities, ethnic and urban groups make similar claims, to such an extent that there is sometimes a sense that everyone on the planet is in state of dislocation. Trying to overcome this situation, we will approach the argument from the point of view of cultural and conceptual evolution, that is, not only as the current trend in sociological description but as a constitutive element of a particular culture.

Western/European universities are still very colonial places, where the type of training, constructed discipline boundaries, languages, and *anagrafe* origins are predetermined elements of any academic career. This book is also about the experience of a non-Western/European academic who, working within a Western/European environment, has always been forced to work in different and non-elected areas ('marginal areas') or, more exactly, who was forced to work in areas that were regarded as a confirmation of foreignness, biased by language, nationality, training, or other elements currently considered as 'local' characteristics of belonging.

Moreover, refusing to accept the *universality* of some languages or conventions (mathematics, science, etc.) – aspects that are part of this 'academic syllabus' – places you at the limit of the academic domain. That is to say, you become and/or you are accepted solely for where you come from, and therefore you will fit in only for that *folkloric* aspect mentioned before. The 'reversal thinking' that we will propose in this book has a long connection with this experience of alienation.

Some authors research in the deposit of archives, in libraries' architecture or in bibliographic whereabouts with great results. But this work is a *search into ideas and projects*, wherever they may be. No one born and bred in a peripheral area of the planet, like Latin America, can ever be a philosopher in the traditional European sense. And that is a good thing. The natural understanding of 'rational' and speculative issues is foreign to non-Europeans. And the difficulties and misunderstandings that this situation produces are endless. Previous local authors, like Enrique Dussel (1934–), complained and indicated limitations resulting from this (see for example Dussel, 1973). We prefer to take a rather more pragmatic and scholarly cynical approach. The result is what you are going to read in the following pages.

Bibliography

Apel, Karl-Otto. *L'*a priori *du corps dans le problème de la connaissance* (Paris: Cerf, 2005).

Canaparo, Claudio. *Muerte y transfiguración de la cultura rioplatense* (Buenos Aires: Zibaldone, 2005).

Dussel, Enrique. *América Latina dependencia y liberación* (Buenos Aires: García Cambeiro, 1973).

Uexküll, Jacob von. *Monde animaux, monde humain* (Paris: Denoël, 1965).

Chapter I
Geography and Philosophy

> You see… in England I am black but back in Brazil I am not… I am white because there are people there who are more 'black' than me.
>
> A. D. S., PhD student, University of Exeter, 2006

> Concepts, as human bodies and objects, suffer and change with altitude, humidity and temperature. What we need is a geo-epistemology able to forecast these weather conditions in our culture.
>
> Edmundo Benaján, 1979

1. The Domain

The concept of *geo-epistemology* that we will postulate here aims to address two main issues: (i) one is the *location of knowledge*, (ii) and the other is the 'épistémo-anthropologique' condition of knowledge and, therefore, the concept of knowledge itself (see for example Apel, 2005: 21). Both issues are obviously related and point to the same conclusion: there is not one unproblematic and simple objective/universal notion of knowledge, despite what many Anglo-Saxon scientific practitioners would like us to believe (see for example Ziman, 2000).

The process of thinking supposes a conceptual system and in the Western tradition 'epistemology' is the name of the domain that considers the appearance, evolution and construction of that system.[1]

1 As far as I know 'epistemology' is a term that was invented as a word by the Scot James Frederick Ferrier (1808–1864) – and used in his *Epistemology or Theory of Knowing* published in 1850 – and developed later as a concept in France by authors like Gaston Bachelard (1884–1962). Some French authors attribute the first use of the word to Bertrand Russell (1872–1970) in his still very interesting *An Essay on the Foundations of Geometry* from 1897.

However, two things make this simple scheme less obvious. The first is the fact that it has proven almost impossible to separate a particular conceptual system from the process and evolution that make it up. This point is made by Jean Piaget (1896–1980) and Ernst von Glasersfeld (1917–) among other Western European authors concerned with the validity of knowledge (see for example Glasersfeld, 1995). Second there is the fact of the *mediation of the body*, that is, the problem of how the conceptual system is affected and/or limited by the use of the body or by the *material dimension* that it represents.[2] This is the problem that in Western terms in the twentieth century Karl-Otto Apel (1922–) has already addressed, and it is also the question at the base of the present contemporary discontinuity between knowledge and perception that has been considered by many authors as the most appropriate definition of 'post-modernity' (see for example Lyotard, 1984 and Virilio, 1991).

Even when considered within these restrictive conditions, this domain of thinking (or conceptual system) remains a European/US property developed mostly in the media environment, in the editorial market and within the corporations that we know as universities. This situation explains why, more than before, a great number of scientific authors are constantly concerned with the idea/notion of knowledge: because knowledge in this sense is validated today less by its epistemological accuracy than by its *public dimension* (see for example the monumental work edited by Latour, 2005). The growing complexity in the understanding of the conceptual system of a particular domain becomes entangled with the issue of the financial dimension of intellectual work in our contemporary societies (see for example Soja, 1989).[3]

2 The notion of 'body' here includes also 'bodies'. In short: the *mediation of the body* is a way of introducing the issue of the physical dimension.

3 The notion of 'entanglement' will be used in this book with the original meaning under which was deployed in the domain of quantum mechanics. Quantum entanglement is a quantum mechanical phenomenon in which the quantum states of two or more objects are linked together so that one object can no longer be adequately described without full mention of its counterpart – even though the individual objects may be spatially separated. As far I know it was Erwin Schrödinger (1887–1961) who initially coined the term 'entanglement' to

What should be added as another limitation to the Western classical idea of the conceptual system is the fact that there is not a *universal* scheme valid to all territories. In the same way that the evaluation of a particular conceptual system cannot be separated from its creation, evolution and history, equally the implementation of a particular analysis regarding a specific culture, community or group, cannot be separated from the land, territory and space in which it has been created, developed, and established. And of course, this *location* can be physical or imaginary or, more precisely, together they constitute one of the contemporary characteristics of knowledge of the present in marginal/peripheral areas of the planet.

This argument of 'universality' can therefore be considered in two ways. First, historically speaking, from the perspective of the history of European philosophy, it could be associated with the classical point of view regarding the perspectives emerging from 'the dispute about the universals' understood as values, ideas or principles that is, about the two perspectives which from Medieval times have been historiographically debated about the possibility and extension of knowledge. From one perspective universals exist only as particular *nomina*, and from another there is a general real dimension that cannot be overcome. Perhaps it can be better understood from this context why many Latin American authors find in European 'nominalism' and in European thinkers like Juan Duns Scotus (1266–1308) – or in his successors (see for example Eco, 1986) – a better understanding of the World-place (*Dasein, habitat*) issue as of the possibility of knowledge from a theoretical point of view ('teoría del conocimiento'). This connection is obviously instituted less because of its viability and historical pertinence than because of its appeal to a strong imaginary presence within the definition of knowledge. But, nonetheless, a connection is also made from the fact that the debate about 'universals' places at its centre the existence of what we can call *das Ding* ('la cosa', *res*, *dinc*, the thing, the *thinging*), defined as the

---

describe this peculiar connection between quantum systems when discussing the famous Einstein-Podolsky-Rosen (EPR) argument in 1935 (Schrödinger, 1935; 555). Although it might sound a bit unorthodox, for the purpose of the argument, the term 'entanglement' will be used not only as noun but also as verb and not always in the passive voice as would normally be expected.

material unit of analysis and discussion (see Canaparo, 2007). 'The thing' (*res*) becomes an inevitable reference within the arguments and a basic component of the philosophical and cultural evolution – the case of Martin Heidegger (1889–1976) is one of the most clear in Europe within the twentieth century (see Heidegger, 1971 and Edwards, 2007). And this explains why the idea of *materialidad* or materialism (see for example Galeano, 1986, 1987a and 1987b) has always been present in cultural and/or philosophical analyses of Latin America (see Canaparo, 2007). 'El pensamiento de la cosa' [the thinking of the thing] is at the limits of twentieth-century Western philosophy and its language (see Heidegger, 1971; and also H. L. Dreyfus/M. A. Wrathall, 2007).

Within the classical approach to the possibility of knowledge, from the perspective of nominalism, universals are not *ante rem* – they are not there before the thing – and nor are they *in re* – they are not in the thing itself. This fact makes it attractive to peripheral areas like Latin America where this *absence of the thing* is almost a regular phenomenon in the naming process of 'world' (immediate environment) and also in the construction of a principle of reality. And this is also why so-called 'methodological nominalism' – that is, a more extreme perspective where the problem is not between abstract and concrete entities but one which assumes that the only existent elements in our construction of the world are solely concrete entities – becomes a very *local* approach to the present understood as a problem. The wide local use of the so-called Frankfurt School's approach (including the works of Walter Benjamin [1892–1940]) to culture, society and modernity is indeed connected with this *materiality*.[4] More recently, the academic popularity of Anglo-Saxon authors like Richard Rorty (1931–) in Latin America has certainly had to do with this aspect. Equally, the relevance of all European philosophers who placed the *linguistic turn* at the centre of their works can be understood for the same reason.

4 We will return upon this aspect of 'materiality' later on in this chapter. Academic works *about* Latin America using the so-called Frankfurt School as a sort of reference are endless. Similar is the case for the works of Benjamin. Perhaps these developments can be better observed, for example, in relation with the journal *Punto de Vista* (1978–2008).

Secondly, we need to consider this 'universality' in relation to the present situation of academies and universities where the writing activities are standardized to the extreme by the editorial and media markets – to such an extent that the so-called 'written style' of an author became an instrument of epistemological control: what is relevant is not the semantic accuracy and pertinence of a particular concept but how it will fit within the 'Guidelines for Authors' of a particular publication or publishing house (see for example Canaparo, 2003 and Collins, 1998). No less important is the fact that this 'universality' has evolved in such a historiographical way that within European culture it is nowadays almost impossible to separate knowledge, science and the written dimension that makes it transmissible and public (see Canaparo, 2003). Basically this means that the distinction between information and knowledge is less and less functional, as some authors have already pointed out (see Lebeau, 2005). Situations that have a direct relation with the extended belief, among Latin American authors and analysts, that understand the media and are in possession of media information become synonymous with being part of what happens in Central-imperial parts of our planet.

The study of peripheral areas of what is understood as Western culture is directly connected with the *location of knowledge* (the first indicated issue) and with the principle of an 'épistémo-anthropologique' condition, present within all definitions/approaches to knowledge (the second indicated issue). The understanding of Latin America as an idea, place or perspective cannot avoid its long-standing connection with Western/European culture and therefore an analysis/commentary is necessary regarding the two above-mentioned aspects – not only for what they could be by themselves (as analytical development) but also for the context in which they are placed and the function they have. We will indicate this analysis/commentary as 'geo-epistemology' (for an extensive use of this concept see Canaparo, forthcoming).

The notion of geo-epistemology will try then to argue that 'Latin America', as our principal leading case and as a peripheral space of Western culture, is less than an 'ideology' or a particular set of folkloric issues and elements, but more a perspective about knowledge, comprehension and writing.

2. The Question of the Location

2.1 Identifying the Local

The question of the location of knowledge refers to the fact that not all conceptual systems can be extracted from the environment in which they were created and conceived.[5] A particular conceptual approach does not need to be systematic or coherent to be considered under this geo-epistemic condition: most concepts emerge in a rather chaotic environment and it is through their use and functionality that they become part of a broader system or scheme. The question of the location of knowledge refers to the impact that (i) cultural evolution, (ii) historiographical perspectives and (iii) public diffusion have in the conception and design of the meaning of concepts themselves.

The River Plate writer and analyst Ezequiel Martínez Estrada (1895–1964), together with the Mexican writer and intellectual Alfonso Reyes (1889–1959), is probably one of the authors who, in the space we call Latin America, has discussed in more varied ways the notion of *local* and its relation to an idea of culture. Even when Martínez Estrada, and indeed Reyes, were thinking in *European terms*, their approach constituted an inevitably twisted version of those so-called European arguments. This *ratio obliqua*, as we will call it, was/is an inevitable deformation and/or degeneration of European perspectives implemented within the colonies. However, neither Martínez Estrada, nor Alfonso Reyes, ever fully discussed the three aforementioned aspects before considering the construction/implementation of a concept of knowledge

5 The major contemporary discrepancy within academic/university publications between the title of a monograph and the contents of that monograph are astonishing – and constitutes a perfect example of this condition. Gérard Genette already noticed this situation in its early days and described it as *l'effet Jupien* (see Genette, 1997). However Genette described the situation as part of an author's paratextual strategy and we believe it is more important than that and therefore connected with the central epistemological problem described here. The problematic distinction between a 'context of production' and a 'context of reception', when referring to a particular concept or theory, is another interesting example of case that can be observed within the colonies' environment.

– and its impact – in connection with its location. However there is something in the way that Martínez Estrada was trying to implement concepts that makes him a *local character*.

And it is this 'something' that needs to be addressed when we refer to Latin America as a colony and as a territory at the margins of Western culture: even with the most extreme version of imperialism in place, the local dimension was never totally dominated by central authorities or institutions. There was always an element of *degeneration*, a twisted and unexpected consequence of the monopolist flux of information. Not so much generated for some ideological or political reasons, not so much a product of some kind of rebellion, these non-linear elements ('alternativos', 'no alienados') within the colonies are produced by degeneration, by a *natural implementation* of the *ratio obliqua*. The anti-cartesian perspectives proposed by some authors as a necessary alternative to the Western/European notion of *cogito*/individual (see for example Mignolo, 2005), ignore this long and degenerated cultural/conceptual evolution within the colonies. Moreover, to believe that some political dimensions ('cooperativas', 'movimientos alternativos', etc.) are going to change the elements of colonialism which, as a long-standing presence have been considered by the local State authorities and intellectuals as the 'real identity' (the cases of Argentina and Mexico are perfect in this sense), is to ignore the evolution of the culture and the concepts within the colonies themselves. Whatever is 'new' out there, on the streets as well as in the institutions, is always forced to *acknowledge* this past sense of identity. Even more importantly, it is forced to acknowledge and become familiar with a lexicon and a set of concepts that are already there and that are at the core of the present sense of community and society. Even more radical is this conceptual situation, where the impact of the *media* in that particular community or society is more important (see for example Miller, 1991).

Rather than being connected with some hypothetical rescue of *aborígen* cultures, the comprehension and modification of colonialism need to go deeper into the analysis of the *ratio obliqua* – by exploring its mechanisms, conditions and consequences. Latin America is a space dominated by a present and contemporary culture which has at its core a strong *media* domain. Any theory or proposal that ignores this issue will lack efficacy and will struggle to make its way within the local market and the public domain.

2.2 Constructing the Distance

The question of the location of knowledge as a product and/or entanglement between Western philosophy and geography, has been addressed, at least in the way that initially concerns us, first by philosophers like Gilles Deleuze and Felix Guattari (1980 and 1991), Hans Blumenberg (2000), Randall Collins (1998), by geographers like Edward W. Soja (1989), Horacio Capel (2001), Jacques Lévy (1999), and David E. Nisbett (2003), then by Western cultural analysts like H. Bhabha (1994) and Slavoj Žižek (2003), and, most recently and more closely to our subject, by authors concerned with the idea/notion of Latin America, such as the Argentinean-born American Walter Mignolo (1934–) and the American academic Ramón Grosfoguel (2006), among others. What makes these authors interesting is not only the above-mentioned connection between two Western academic areas of study but also the fact that, each within their own scheme, they have been thinking about the limits or, more precisely, they are consciousnessly thinking at the limits and borders of Western culture.

The connection between thinking geographically about the borders of a territory and the speculation about the limits of our knowledge is an issue that goes back to the great cartographic development initiated in Europe after the 'discovery' of America and which reaches its climax in the seventeenth century. However, in almost all cases, this was a *philosophical issue for cartographers* and not the other way around. It was only in the twentieth century, with the *crisis* of the notions of representation, measure and matter, that Western/European philosophers were forced to think outside the historiographical tradition and in a new form to approach space. Indeed, this situation also partially explains why the so-called *linguistic turn* was so strong and dominated almost the entire philosophical Western *escenario* regarding knowledge and the theory of knowledge during the twentieth century.

The pioneering works of Henri Bergson (1859–1941), Niels Bohr (1885–1962) and the already-mentioned Gaston Bachelard, among other European thinkers, were guided by this concern. It is necessary to acknowledge that the conceptual problems faced by them forced the dissolution of the traditional boundaries of what was considered as philosophy, especially, as we mentioned earlier, due to the increasing legitimacy of *scientific knowledge*.

Thus, the problem of space was then threefold: (i) first, as a traditional-Kantian concept to be addressed, (ii) second, as a problem of the philosophical historiography itself, and, (iii) third, as a cultural issue within an environment that was dominated by the *visual* and by an increasing/expanding use of cartographic principles and approaches.

When Gilles Deleuze (1925–1995) and Felix Guattari (1930–1992), in 1980 and later, confronted the issue of *territory* in philosophical terms, the *milieu* was ready for their hypothesis: the limits of philosophy as a speculative/language-orientated activity cannot be separated from the definition of space itself – land, desert, territory, etc. (see for example Deleuze/Guattari, 1980 and 1991). The main concern of Western academic philosophy became, towards the end of the twentieth century, the establishment/design of *distances*, or more precisely, the problem of how we define our concepts – or conceptual system/scheme – in relation to historiography and in relation to other theories and concepts.

Within this context, and not by chance, many authors of the colonial territories – or of the borders of the Empire – were attracted to several of these Western/European thinkers. They thought, that by becoming thinkers of the borders and limits, they would help to establish a non-European view about space. This, of course, was a short-sighted judgement as the concepts proposed by these authors might have helped but only in relation to the broken-down Western/European historiography and its institutional legitimacy. The *biografismo* (biographism: the use of biographical narrations as the main explanation of ideas), which dominates the use of sources and the conception of history of science and philosophy among academic authors in Latin America, can be explained by this already existing situation.

The development in the *sede filosófica* [philosophical domain] of several issues about space – as a key speculative ground – forced Western authors to consider the meaning, viability and use of the very idea of *concept* itself. And that is why the so-called radical constructivism – as in von Foerster's notion of 'cybernetics' – becomes, situated at the limits of what is possible in intellectual and institutional terms, a powerful instrument. It is a theory/approach that entangles (i) the construction of space, (ii) the concept of knowledge, and (iii) the establishment of a subjectivity at the same time. Therefore, the implicit 'constructivism' of these above-mentioned Western/European authors – despite in some cases their own views – constitutes the climax for what we can describe as the questioning of the conceptual viability and legitimacy of Western/European historiography. The fact that many European thinkers presented themselves publicly as *foreigners* with regard to their own origins and traditions is indeed a non-equivocal signal of this condition. Equally, of course, many colonial/marginal authors started to consider themselves almost as Western/European. This *confusion* between what is at the 'centre' (Empire, imperial) and what is at the 'borders' (colonies, periphery), or, more precisely, how within the 'centre' there are some peripheral areas, and, on the contrary, how in the periphery there are parts that are 'central', is what we experience today.

Western/European authors, as well as colonial-born and/or educated artists and authors, have confronted the burden/burning of boundaries by constructing distances (i) between central and peripheral parts of the Empire, (ii) between physical geography and the imaginary/imagination of land, (iii) between intellectual/philosophical limits/borders and geographical constraints, and (iv) between the limits of language and the epistemic condition of the perception of the physical world.[6] This general burden/burning process partially brings us to another important and specific characteristic of the Latin American space. This is connected with the fact that the original mentioned relationship between thinking geographically about space and speculating about the notion of knowledge is entangled with the idea/concept of imaginary,

6 The burden/burning condition of the boundaries refers to the fact that boundaries operate with a double characteristic: as weight ('peso', 'pasado', 'carga') and also as something that is burning ('quema', 'destrucción'). The condition of boundaries, as introduced here and we will see later in the book, does not only refer to a number of characteristics, such as, for example, instability. It refers also to the fact that boundaries compel people to do things, to behave in a particular way, to perceive in a particular form.

imaginatio and *imaginación* – a condition we have already explored in *El imaginario Patagonia* (see Canaparo, forthcoming). The fact that designing and naming space brings the needs to re-formulate the idea of imaginary/imagination should not be a surprise.

3. The Question of Comprehension

3.1 Exploring the Limits: La frontera

Understanding understanding ('comprehension') has been described, at the limits of Western culture, as the only way to have a conceptually appropriate deployment of contemporary philosophy (see von Foerster, 1981 and 2003). In general, the impossibility of conceiving 'knowledge' without an idea of education makes the concept of 'knowledge' dependent on the meaning ('acepción') of learning. At the limits of the Western/European way of thinking and culture the concept of knowledge drives us to a more socially orientated concept of education that is directly connected to an idea of learning. And it is the regular interactive relationship between the concept of knowledge and the idea of learning that generates the possibility of what we call 'comprehension'. This is also why the so-called 'radical constructivism' is, we believe, clearly understood and describes so well the case of the Western/European way of thinking – even conservative and traditional US philosophers agree on the consideration of this pioneering perspective of the constructivist approach (see for example Lakoff/Johnson, 1999). What was (and still is) crucial as well within this approach is the fact that the notions of *communication* and *circulation* were placed at the core of the very idea of knowledge.

More precisely, so-called 'radical constructivism' addressed a general and common question: a description or definition of comprehension ('How do we learn?', 'what is the meaning of learning?', 'how do we understand?', 'what is the meaning of understanding?') which is the first step to discussing and/or approaching an idea of knowledge in Western cultures. Comprehension ('understanding understanding') can only be achieved in contemporary Western culture by addressing

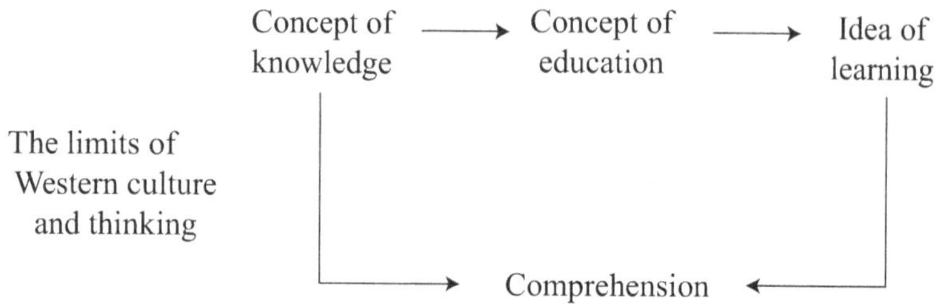

FIGURE 1.1 Basic process of circulation and communication at the limits of Western culture and in relation to the idea of knowledge.

communication and the circulation of information as a basic ordinary activity, that is, conceiving knowledge as a communicative and always-in-movement object and/or device. Comprehension ('understanding understanding') is the real guiding element in the Western notion of knowledge. That is why, for example, in the philosophical domain of academies and universities the theory of knowledge – or different versions of it, as for example, 'sociology of knowledge' – almost completely absorbed all the activities during the twentieth century as the most problematic area. Furthermore, that is also why the distinction between 'explanation' and 'understanding' has been used as a *distinction* that characterizes the superposition ('hipóstasis', 'superposición') between knowledge and scientific knowledge. Moreover, it can explain why the only total ('holistic') and comprehensive definition of 'culture' towards the end of the twentieth century in Europe has been dominated by the notion of 'science' or 'scientific knowledge' (see for example von Wright, 1971; and also the instructive Apel, 1984).

Within this context of understanding understanding ('comprehension') one main element is the acknowledgment of the limits of concepts and the activity of thinking. The works of Karl-Otto Apel 'Technognomie' ('Technognomie – eine erkenntnisanthropologische Kategorie') and 'L'*a priori* du corps dans le problème de la connaissance' ('Das Leibapriori der Erkenntnis. Eine erkenntnisanthropologische Betrachtung im Anschluß an Leibnizens Monaden-Lehre'), published in 1958 and 1963, constitute an unexpected, short and very effective description of an epistemological question which, as yet, has not been much discussed: is it possible to postulate acquired pre-conditions to our knowledge?[7] The philosophical historical approach of Apel is less important than one aspect/distinction that comes up several times in his two articles:

> Une disjonction complète de l'empirique et de l'apriorique dans notre connaissance: d'un côté, il s'agit de tout ce qui peut être expérimenté en général, de l'autre, des conditions de possibilité de l'expérience, inaccessibles à l'expérience. (Apel, 2005: 10)

In other words, this is the problem of incommensurability between perception and knowledge that became considered as the condition of our understanding after the disappearance of modernity's values (see for example Lyotard, 1988). More specifically, this is the question that, under the name of 'present' or 'événement' has been discussed by authors like Jacques Derrida (1930–2004) and Alain Badiou (1937–) (see for example Derrida, 1978 and Badiou, 2005). Or, from a broader perspective, from authors who analyze the notion of culture or technology in contemporary European terms, like Paul Virilio (1932–) and Slavoj Žižek (1949–), discussing the impact of images, war conflicts and media development in relation to our perceptual capabilities and regarding our speculative activities towards the past and the future.

Apel addresses an issue that has been considered before in broad terms by Friedrich Nietzsche (1844–1900) (see for example 1962 and 1974) but always considered to be *marginal* to 'real' philosophical problems:

> La division platonicienne et cartésienne des sphères requiert ce rapport du sujet et de l'objet, mais le concept d'une représentation perspective du monde et, par conséquent, d'une représentation toujours déjà médiatisée par le corps interdit cette division parce qu'elle ne peut pas être appliquée au rapport du sujet de la représentation et de son propre corps sans *regressus ad infinitum*. (Apel, 2005: 15)

7 Of course, in the history of Western philosophy this question has been asked on several occasions. What is new is the fact that the question has not been discussed extensively in the light of a 'post-rational philosophy' (R. Rorty) or under the conditions of a 'post-industrial society' (Z. Bauman).

And further:

> La distinction sujet-objet et la disjonction correspondante de l'apriorique et de l'empirique dans la macrophysique classique aussi bien que dans la théorie de la connaissance traditionnelle de la philosophie représentent, à mon avis, une idéalisation de l'état de choses qui n'est adaptée qu'au seul cas particulier de la mise à disposition technique des objects du monde ambiant. (Apel, 2005: 34)

Besides the historical issue regarding philosophical theories of knowledge, here we can see (i) how the way to associate technology and science is already there, and (ii) how the question of technological objects and devices will conform and dominate the idea of a *real environment* (see for example Lebeau, 2005).

The historical, philosophical and historiographical approach of Apel prevents him from acknowledging, for example, the works of Jean Piaget and the fact that Piaget, among others, has been working on exactly this problem of the 'body' and the constitution of knowledge in human terms (see for example Piaget, 1971). Furthermore, authors like Ernst von Glasersfeld have taken Piaget's theories one step further and entangled them with the question of the conditions of scientific knowledge. However, the implicit 'constructivist' approach of Apel is somehow present, for example, in the reference to the works of J. V. Uexküll (1864–1944) and in its comment about 'sa doctrine épistémo-biologique de l'*Umwelt*' (Apel, 2005: 23).[8]

It is interesting that the alternative proposed by Apel to face this epistemological problem is to consider Einstein's theory of relativity as a driving force in considering knowledge and a theory of knowledge in philosophical terms. I said this is interesting for two reasons: first, because Apel relies on a scientific theory to justify a philosophical approach and, second, because outside the theory of relativity, Einstein's notion of physics and the 'world' is a very traditional and Newtonian

8 Of course, this 'constructivism' relies on more than one reference to Uexküll. Generally it can be found in the idea that the we cannot have a notion about a particular object without the observer that makes that object what it is (see Apel, 2005: 65). Or also on the assumption that the reality of matter is something that is not available to the perception and therefore invisible to human being (see Apel, 2005: 53 ff.).

one (see for example Pais, 1982). Apel makes a reference to Niels Bohr (1885–1962), but does not pursue this connection further that, contrary to Einstein's perspective, could take him much deeper in this question of the conditions and limitations of knowledge.

In any case, Apel did not lose sight of the central problem and asked the crucial question:

> Dans quelle mesure, ou jusqu'à quel point, l'homme est-il effectivement vérifiable à partir d'une position excentrique, c'est-à-dire des théories au sein desquelles son propre point de vue de la représentation du monde peut être rendu disponible de manière à ce qu'il puisse constituer une donnée empirique possible? (Apel, 2005: 18)

It is well known that this question has been given a negative answer (see for example Piaget, 1972) or has been considered as a question without an answer, and thus as a diagnosis rather than an interrogation (see for example Sartre, 1958). More importantly, Apel already predicts here the grounds of the future discussion about scientific knowledge that would be brought by the so-called 'Science Studies' years later, that is, (i) the non-objective role of the observer (see for example Glasersfeld, 1995 and 1987), (ii) the impossibility of representation as primary reality (see for example Enaudeau, 1999) and (iii) the always relative notion of the empirical approach or data (see for example Feyerabend, 1981).

The issues of speed and simultaneity, that nowadays seem to be very common to us (see for example Virilio, 1991 and 2007), were also presented by Apel in very simple but effective terms:

> Ainsi, la divergence apparemment paradoxale du constat de simultanéité de systèmes référentiels qui sont en mouvement les uns par rapport aux autres repose sur la mediation sensible du concept de 'présent', ou la 'simultanéité' par l''être-au-monde' corporel du sujet connaissant. (Apel, 2005: 25)

The aspect/problem of the 'present' is then not only how we acknowledge what is happening or how we can name the pure becoming ('devenir'), it is also, and no less importantly, the fact that our notion of knowledge is connected right from the beginning with the *possibility* of knowledge itself. We can understand why Apel, like many other European philosophers of the twentieth century, would go in the direction of *language* to construct his theory.

Apel has left us – and that is why we mention his works here – with the notion of *concept 'épistémo-anthropologique'* (Apel, 2005: 21), which can be characterized following the brief discussion established above and which will be very useful in addressing the issue of knowledge in peripheral communities and regions like the space of Latin America. Apel calls his research and approach, grounded in this *concept 'épistémo-anthropologique'*, as the 'principle of *Technognomie*':

> En partant du problème de la technognomie, la question d'une anthropologie de la connaissance comme programme d'investigation d'une 'raison concrète' qui se rapporte au corps, par opposition à la théorie de la connaissance de la conscience pure et réflexive et à ses structures immanentes, fondée par Descartes et Kant (à partir de Platon), devient implicitement possible. (Apel, 2005: 47)

And although Apel never extensively developed this approach, its pertinence and effectiveness still reaches us, especially when we try to explore the borders and limits of the European/Western notion of knowledge that forms our current global sense of science and culture. And, even more importantly, Apel did establish a 'double' sense of the 'épistémo-anthropologique' principle: (i) the issue refers on one side to a theory of perception and to the idea of experience (individual level, process of subjectivization), but also, (ii) on the other side to a theory of knowledge and to the fact that the foundations of any community, institution, group or society includes without exception *language* (collective level, process of knowing) (see, for example, Apel, 1975).

3.2 Knowing/Living in the Borders

Understanding Understanding. Essay on Cybernetics and Cognition by Heinz von Foerster (1911–2002), published in 2003, is a collection of essays that covers the period from 1959 to 1993. In this work, as in others before, Foerster postulates that the perception of limits ('the unthinkable') is *connatural* with the activity of creating concepts ('conceptualizing') to understand our immediate environment. The notions of 'cybernetics', 'self-organizing systems', 'reality', 'epistemology of leaving things', and 'communication', among others, were a set that within this book established a *borderland* of thinking not as an exceptional situation but as a normal procedure. Going to the limits of what could be thought or named has been the norm within Western philosophical historiography. And, what I find interesting within 'radical constructivism' as formulated by Foerster – in line with the works of Ernst von Glasersfeld, Francisco Varela (1946–2001), Humberto Maturana (1928–) and Nikhlas Luhmann (1927–1998) – is the fact that this limit becomes *the normal environment of thinking*: 'living' in the borders is, according to these authors, the only way to avoid the mainstream in conceptual terms – what we can call the Western monopolic vision of philosophy, that is, a sort of combination of nineteenth-century European realism with seventeenth-century European natural philosophy (for an extended discussion about this realism-naturalism and its scientific results see Canaparo, 2003).

The popular version of cybernetics refers to an entanglement between man and machine. However, a more accurate description, following authors like Heinz von Foerster (2003), should place us in relation to a most interesting and complete domain, that is, towards a notion where it is the whole idea of *immediate environment* that is under analysis. This was also one of the basic points of Norbert Wiener's (1894–1964) seminal idea of cybernetics (see for example Wiener, 1961). It is in this sense that we consider cybernetics, within the twentieth century, as one of the limits of Western culture where the spatial consideration of it and the conceptual analysis attached to it go together. The other domain, also situated at the limit of what is possible and thinkable within Western culture, that we analyzed elsewhere (see Canaparo, 2003), is the field of quantum mechanics in the interpretation of Niels Bohr or, more precisely, the conflicts and arguments arising from the creation and development of quantum mechanics as a scientific area.

We have two aspects, one somehow generated from technologies and machinery ('cybernetics') and the other from an extreme version of science and scientificism ('quantum mechanics'). Being on the borders of what has been considered as Western culture, they show the close connection that exists between (i) the definition of matter, (ii) the establishment of a concept of invisibility, (iii) the transformation of science in production of technology, (iv) the determination of a notion of scientific space as a new universal value, (v) and the creation of a new notion of *geography* that is governed by images of what is not visible-perceptible – the new maps – and by concepts and scales which are not measurable outside lab experiments and developments. This new Western *geography* has already been analyzed-exhibited physically (see for example Latour/Woolgar, 1979; and Galison/Thompson, 1999), mathematically (see for example Thom, 1983) and in an imaginary way (Montgomery, 1999 and Whatmore, 2002). But, more importantly, this new geography is a confirmation of the scientific domination of the Western notion of culture and the establishment and determination of visual elements as probe/proof of existence. Western reality, in spatial terms, paradoxically for an historical period where matter became invisible, can only be 'scientific' and 'visual'.

And if this explanation is necessary it is because, the more Western/European culture approaches the limits of its own possibilities and conceptions, the more some authors believe there is a sort of 'crisis' in Western/European terms or, similarly, that there are more *insterstícios* [interstices, cracks] for non-Western/European thinkers to consider the present as a way to change the conditions of knowledge. There could be nothing more illusory. The colonial/imperial domain of knowledge veers nowadays towards the idea of the market and a new formulation of what is considered as the public sphere, areas in which peripheral cultures do not have a chance to establish alternative views. However, this going to the extreme financial and sociological formulation of Western cultures must not be confused with the 'living at the borders' proposed by radical constructivism and which is somehow a reality within Latin American space.

What we can consider as the natural environment of the space indicated as Latin America is that there is a *continuum* between having been/being at the borders of the Empire and thinking/writing at the limits of the Empire. Nothing defines Latin America more than a *continuum* between a physical spatial location and intellectual written location, as well as between an imaginary location and a market/public location. The regular local misunderstanding/confusion, for example, between historiography and the publishing market is a historical result of this situation. The question of Latin America as a given space is not – as many nineteenth-century authors, politicians and intellectuals thought – that it is situated 'beyond the bounds of Civilization'. It is much more difficult than that. Latin America is placed in a grey no man's land that can be considered as a *frontera* in its own totality. In *Muerte y transfiguración de la cultura rioplatense* (2005) and further in *El imaginario Patagonia* (forthcoming) we have already developed this perspective in the context of the River Plate and Patagonia. This is an area in which, because of its constant development against an imaginary borderline of civilization and against a physical borderline of country/nation, the local culture became an all-exhibited space dominated, on one hand, by many of the archetypical nineteenth-century elements of *frontera* (as division, as difference and distinction, as a limit or border, as definition of foreignness, etc.), and, on the other hand, by many of the elements generated by the *sui generis* post-industrial societies (as a transition and *continuum* condition, as a necessary element to define the 'local', etc.) that the great majority of the big cities of the planet had become towards the end of the twentieth century. This characterization can be made extensively of the space of Latin America as such.

And this *situación de frontera* (borderline/frontier condition), as previously suggested, is one of the reasons why the notion of space is central to Latin America, but no less important is the fact that it makes it much more difficult to think about the influences of colonialism/imperialism as they are directly entangled within this *zona gris* (borderline condition). The arguments of the 'indigenismo', as discussed by José Vasconcelos (1882–1959) (see for example, 1976), among others, the ideas on the 'theory of liberation' from Gustavo Gutiérrez (1928–) among others (see for example Gutiérrez, 1988), of the 'revolution' of education from Paulo Freire (1921–1997) and others (see for example Freire, 2000), and the perspective of 'delinking' from Walter Mignolo and others (see for example Mignolo, 2005), share a proposed direction. Under these above-mentioned conditions, they superimpose ('superponiendo, superponiéndose') the way in which the political ruling class and its 'criollo' intellectual thinkers – who used the idea of 'modernidad' as a driving element of analysis (see for example Sarlo, 1988) – have considered the problem of 'identity' and nationhood. The element to consider here in order to establish a *difference* between theories, people and ideas is less a narrative content or argumentative principle than a way in which these narrations are distributed at what we describe as the geo-epistemic level. This is why it became common to use a *time-scale* parameter for analytical purposes (chronologies, dates, historiographies, etc.). In the context of a general time-scale parameter, either philosophical or political reasons make all these varied approaches pretty similar when it comes to deciding what is conceptually different and why. And, more importantly, the time-scale parameter proved to be very useful and reliable, in colonial terms, at the moment of justifying why a particular set of concepts is more effective and/or legitimate than another.

4. Spatial Thinking

Both questions/approaches – the question of the location of knowledge and the question of the comprehension of it – arise from a *spatial* consideration of the speculative domain. Chronologies and, more importantly, the concept of *historización* itself (*Geschichtlichkeit*) constitute an imperial perspective, that is, as far as our arguments are concerned, a dominant perspective that prevails within North American and European academic institutions and publishing markets. The axial place that the works of Martin Heidegger had during the twentieth century among European and US thinkers and authors is, in our view, a clear consequence of this constant *historización* of the activity of thinking itself (see for example Lampert, 1974; Bartky, 1979; and Sandbothe, 1999).

The first question arises from the notion of knowledge, or more precisely, from the relevance of the notion of science and understanding of historiographical philosophy that Western/European culture implements at the moment of defining knowledge. The second question arises from the definition of the individual in speculative terms, that is, from what has been indicated as the formation/construction of the *subjectivity* (see for example Foucault, 1990 and 2005) in relation to an intellectual/

Area	Geometry	Geography	*Imaginatio*
Concept/idea	Philosophy	Culture	Vision/Eye

TABLE 1.1. The areas and concepts of thinking at the borders of Western/European culture.

educative activity.[9] While the first question brings a problem of cultural heritage and language evolution, the second poses a more basic bio-epistemic question in relation to the limits and conditions of human knowledge.

As indicated before, these two questions are developed at the limits of the possible and at the borders of what has already been established within Western/European culture. In fact, a great number of Western/European artists, thinkers and philosophers are authors of the limits and/or of what is possible but not yet 'named'. Is it not for reasons of serendipity that those Western/European authors are among the most popular within the colonies and within the space of Latin America – from J.-P. Sartre (1905–1980) to M. Foucault (1926–1984), from C. S. Pierce (1835–1914) to U. Eco (1935–), from W. S. Burroughs (1914–1997) to D. Copeland (1961–), from S. Žižek to G. Steiner (1929–), the list could be endless. Therefore, for this double reason, attending to the Western/European situation or looking at the colonies' conditions, exploring these limits and borders of Western/European culture and how they might have influenced the evolution of the colonies themselves seems to be the inevitable path to follow.

And, given this condition, trying to establish a spatial point of view in regard to 'Latin America', is, as indicated above, necessary to avoid the 'colonial historicism' mentioned above which constitutes the *motor* and driving force of what has been understood as European/Western culture – an understanding that also occurs at the border and

9 This is what W. Mignolo in his work about the idea of Latin America calls 'egology', in referring to the historiographical dimension of the notion of unity, autonomy and individuality argued in the *Discours de la Méthode* (1637) by René Descartes (1596–1650).

limits of twentieth-century colonialism/imperialism and which facilitates the regular historiographical justification/legitimation of such an idea of culture. Because the problem with historicism, naturalism and the dominant philosophical *historicidad* is that the colonies are not defined by local elements or situations but by interpretations, values or events coming from the 'central' areas of the Empire or, more precisely, coming from a world-view that conceives a centre/periphery relation.[10] The great paradox of nineteenth-century Latin American intellectuals' and politicians' thinking was that, the more they searched for a Latin America identity grounded in an Enlightenment perspective, the more they travelled towards emptiness/nothingness (for a discussion about this issue see Canaparo, forthcoming).

Thinking spatially then means at least being forced to confront these two aspects/questions which, according with the view expressed here, are connected by a triple approach or, more exactly, by the consideration of three concepts/areas and their consequences and developments. The concept/areas of geometry, geography and *imaginatio* are those three fields that will allow us to define 'thinking' in a more accurate way in relation to the space that we call Latin America ('colonial space') and, consequently, in direct connection with the epistemic condition indicated above. The concept of 'geometry' will bring us the issue of philosophy, its status and definition; the concept of 'geography' will bring us the definition and status of culture; and the concept of 'imaginatio' will bring us the definition and function of vision and visuality (see Table 1.1).

4.1 From Geometry to Philosophy

Western/European philosophy is a *historiographical phenomenon* that uses writing as a dominant device and tool for change. The historical study of this historiographical domain is a complete waste of time for authors living within the colonies and/or on the borders of Western

10 That is why a constant and unavoidable aspect of *translation* is always present within the colonies. We will discuss this issue in Chapter 3.

culture. However, one cannot say the same with regard to the study of this historiographical domain itself as a way of understanding how Western/European approaches have set up their limits and how the concepts and theories are working in relation to them.

One model of this historiography is indeed the French Jewish Algerian-born philosopher Jacques Derrida commenting on the *archi-texte* perspective of Plato's works as well as translating and commenting on Edmund Husserl's (1859–1938) *Origin of Geometry* (*Die Frage nach dem Ursprung der Geometrie als intentional-historisches Problem*, 1939) and suggesting that, at the origin of Western philosophy, as well as *writing* (see Derrida, 1976), we have *geometry* (see Derrida, 1989 and 2003). Thinking is to think spatially even when the Western philosophical tradition seems to point, according to Kant (1724–1804) and Hegel (1770–1831), in the opposite direction, that is, in the historical-orientated pattern (see for example Rorty, 1980). The dual foreign/European condition of Derrida's position makes his interpretation possible. And this is probably one of the reasons why he has become one of the most quoted and commented-on European authors of the last twenty years within Latin America. It cannot be denied that his ideas about philosophy as activity and/or area at the *marges* ['margin'] have strong connections with the way some authors have considered Latin American as a marginal space (see for example García Canclini, 1992; and also Martín-Barbero, 2001). And this marginal space is at the same time conceptual (see Derrida, 1972: Part 3), physical and/or architectural (see Derrida, 1972: Part 4), and intellectual and/or imaginary (see Derrida, 1972: Part 5). This interpretation of philosophy from Derrida is an attempt to reverse the dominant philosophical historiography by reading the philosophical tradition itself – an epistemological attempt and strategy that we will allude to further on in Chapter 5 under the heading of *reversal thinking*.

No less relevant is the fact that, as can be inferred from Derrida's book (1972), there is an historiographical pathway regarding the Western concept of philosophy that appeals (within the present and in the past) to the territories of the 'colonies'. This pathway starts from the fact that within the Greek world 'philosophy' was born in the Ionian colonies of Athens ('colonias Jónicas') and that since then an important number of authors and thinkers that have attempted to define 'philosophy' have somehow been working at the *marges*, as defined by Derrida. From St Augustine (354–430) to Spinoza (1632–1677), from Nietzsche to Camus (1913–1960) as well as from Freud (1856–1939) to Feyerabend (1924–1994), a significant number of authors working on changes to the Western concept of philosophy came from communities outside Europe or were isolated within the European culture or were emigrants from its own immediate culture.[11]

Euclides of Alexandria (325–265 BC) was an Egyptian author who lived in a Greek Empire and is well known for his foundational treatise on geometry known as *The Elements*. Some authors associated Euclides of Alexandria with Euclides of Megara (450–380 BC), believing them to be the same historical character. Euclides of Megara, born in a colony of the Greek Empire, was a follower of Parmanides' theories and a disciple of Socrates according to Plato's *Fedón*. His apparent preference for solving rhetorical disputes and interpretations using the *reductio ad absurdum* gave him, according to historians, the ideal mentality for the development of geometry, a fact which also explains the connection between the two Euclides. From a peripheral situation towards a definition of 'philosophy' and towards a formulation of geometry as total spatial dimension: here is the origin of the Western pathway. And it is to these roots that René Descartes (1596–1650) also refers. It is not by chance that Descartes, considered to be the founder author of modern Western philosophy, is also the inventor of modern geometry under the name of 'analytic geometry'. In this way Descartes re-defined the double Euclidean tradition. As the inventor of the so-called 'Cartesian coordinate system', Descartes founded analytic geometry, that bridge between algebra and geometry crucial to the invention of calculus and analysis in mathematical terms. No less importantly, Descartes established for the first time a sequence that points in the direction of one *Principia philosophiae* (from 1644, and also *Meditationes de prima philosophia* from 1641), towards an idea of 'totality' in geometrical/

11 The very pertinent historiographical definition of Marx, Nietzsche and Freud as the *maestros de la sospecha* ('masters of the suspicious'), developed by Georg Gadamer (1900–2002) and Paul Ricoeur (1913–2005), is indeed a highly pertinent perspective of this situation (see for example Ricoeur, 1970).

spatial terms (*La Géométrie* from 1637) and then to a notion/concept of the World (*Le Monde* from 1630–1633). Descartes complemented Euclides' double tradition and his historiography. The diffusion of Copernicus' (1473–1543) theories at the same time that Descartes was writing, made the notion of space a crucial element to understanding the *invisibility* of the universe which astronomy postulated implicitly.[12] Here, then, is the beginning of a 'scientific domain' that monopolized the notion of knowledge in the Western/European domain during the twentieth century.

And this situation can be considered an important factor in explaining why Derrida went back to Plato to 'deconstruct' this Western perspective and, rather than be in opposition to it, just expanded it as a means of having a better understanding of the conceptual consequences ('epistemology') of such a historiographical background. The relationship between space, thinking and colonial territory runs through the Derridian notion of philosophy. The translation of Husserl's work *Origin of Geometry* (see Husserl, 1962) as well his comments about it (see Derrida, 2003) are a natural development in our view of this conceptual perspective of what philosophy should be in Western culture, that is, towards a concept of philosophy that tends to become tied up with the periphery of the Western/European domain. 'Deconstructing' philosophical activity should be interpreted here in the light of the 'paradigm of conviviality' postulated for Latin America's own development, by, among others, Walter D. Mignolo (see Mignolo, 2005).

The definition *as philosophy* of the activity of thinking in Western/European tradition is something that paradoxically has become the majority approach for colonial territories or developments – in physical and/or imaginary terms. Jean Pierre Faye (1925–), in his stimulating *Qu'est-ce que la philosophie?* (1997), discusses extensively this

12 The absence then of a distinction between 'astronomy' and 'astrology' can be considered as the open door for incorporating philosophical thoughts into it as well: 'Up to the late sixteenth century, the astrologer is mainly a philosopher who interprets the motions of the heavenly bodies and the laws of nature. He is both an astronomer and a physicist, sometimes he is also a physician, and he considers himself a follower of Ptolemy, Galen and Aristotle' (Bezza, 1986: 2).

condition of limit, travelling and foreignness, of the Western activity of thinking indicated as 'philosophy'. As we will see later, a similar perspective but different reasons can be attributed to G. Deleuze's and F. Guattari's *Qu'est-ce que la Philosophie?* (1991). Thus, the ability *to see* space, to formulate it, to imagine it, is the basic foundation of what has been understood as 'philosophy' in conceptual terms, and it is to this pathway within Western/European historiography that Latin America as a peripheral territory, can be referred.

If 'philosophy' is the *ars* of travelling words, concepts, languages and cultures, that is, considered as the 'voyage de la pensée' (Faye, 1997: 170), then the relation between geometry – understood as a set of postulates that aim to give a complete/total image of space – and seeing/thinking the space is direct and immediate. And it is also for this reason that 'Latin American' is connected to this Western/European dimension, because the relationships between travel, imagining, writing, mapping and thinking, is at the core of what 'Latin American' means within contemporary culture. It is also the reason why a similar *itinerario* has been designed by academic authors within Latin American historiography that goes from a foundational theory about space – see for example Francisco Romero's (1891–1962) *Ubicación del hombre* (1957) – to a more recent multicultural urban complex reality – see for example Carlos Monsivais' (1938–) *Los rituales del caos* (2005). The 'colonial' ideas of Romero can in this sense be associated with the 'chaotic' and 'apocalyptic' interpretations of Monsivais or, even more common in many other cases, the interpretation of Monsivais follows a conceptual evolution from Romero's perspective, that is, from colonialist environment to imperialistic dominance, and from modernity to a set of post-industrial consequences.

4.2 From Geography to Culture

Eratosthenes of Cyrene (276–194 BC) was born in a North African marginal town in Egypt which was still a colony of the Greek Empire. This *double* peripheral condition is interesting when we consider the ideas of *library* and *geography* with which, at the moment of their Western conception, Eratosthenes was connected. From a historiographical point of view, Eratosthenes constitutes the origin of a planetary idea of geography that is better expressed through a determined space conceivable as a library. And it is the evolution of this idea of library that, after the Renaissance period, would give place to a renovation in cartography and in the institutionalization of knowledge through the creation of the first European universities. And within the spatial function of the library, on its evolution, the motto *mobilis in mobili* from Captain Nemo's *Nautilus* – in Verne's well-known book – is indeed one of the most elaborate perspectives of the European expansionism and directly related with this idea of a *close environment as a maximum expression of imagination and travel*. The conquest of the 'overseas domains' *per* the imagination constitutes the climax of Western/European colonialism and the open door to the abstract control of imperialism which, in its more common formulation, will be called capitalism.

What is also interesting in this historiographical context is the aforementioned *double* link: Eratosthenes was on the periphery in relation to Egypt and Egypt was on the periphery in relation to Athens. When Eratosthenes went to study in Athens – as it is believed he did – he was exercising this *double peripheral perspective*. And the fact that he was concerned about the idea of space – arithmetic, geometry, geodesy – at the same time as with the idea of knowledge – his 'holistic' perspective – is in our view connected with this peripheral condition. And if we consider this historiographical account, as many authors do, then the connection with the peripheral space of Latin America seems to be direct: from Paul Groussac (1848–1929) – as *exemplum* of the first Librarian of the colonies (see Groussac, 1901 and 1967) – to Guillermo Kuitca (1962–) – as example of the last geographer (see for example Canaparo, 2000) – there is a link/evolution that can be traced in a similar kind of way to the way Eratosthenes' notion of geography can be connected with his idea of how a library contains and designs the idea of knowledge.[13] For Eratosthenes 'geography' was basically *cartography*, that is, 'mapping' what is not visible to the eye or, more

13 This is an example that only covers the River Plate area. Similar perspectives can at least be indicated in regard to other Latin American areas such Mexico, Brazil and Chile.

precisely, establishing possible ways to design the space that is not directly visible to the eye. Similarly, a library was (i) a physical and (ii) an imaginary space that by acting together created an *architecture/ architexte* for a set of information, principles and items that we associate under the term of 'knowledge'.

As discussed earlier, the notion of geography itself is then also entangled, to keep the same example, within the River Plate context, with the idea of library (see for example Canaparo, 2005). In ancient European history – as in the 'colonial' period of Latin America – the first geographers behaved like librarians. In this sense, Borges' (1899–1986) theories and narrations are not *rara avis* within the local environment, just as the cases of authors like George Steiner and Umberto Eco are not exceptional in the Western/European context. The activity of thinking, considered as a way to define/characterize a particular environment and/or space, was fundamental to determining the colonial dimension or, more precisely, to determining what was 'colonial' and what was not. There are even some extreme cases, always within the River Plate area, where what was considered in the late nineteenth century as the *model* of a librarian was a French intellectual transplanted to Buenos Aires.

Travellers, philosophers, outsiders and slaves were always at the origins of what the Western/European culture considered *geography*. More than 'the discipline of/about space', geography has always been in colonial lands a *discipline of thought* and in this sense there is a very close relationship between (i) defining culture, (ii) imagining the space (urbanism, architecture) and (iii) establishing cultural traditions via historiography. This 'cartographic' activity – which became geography because the notion of culture defined by the library was understood as a symbolic and psychological space of knowledge – can be tracked down in almost every colonial community in the Latin American space. No less importantly, it can be also considered at the conceptual *limits* of Western/European traditions of thought – a situation that has been analyzed and discussed extensively by the already-quoted Jacques Derrida (see Derrida, 1972, 1978 and 1995).

The contemporary works of the philosopher Edmund S. Casey (1939–), in particular his idea of 'place' and 'place-world', bring to light within the Western/European historiography this ancient domain of Eratosthenes. Casey fully reinstalls the relevance of space in determining the philosophical domain and the conceptual analytical/descriptive sets/systems within Western/European culture. According to Casey *questions* and *problems* are not determined by a particular subject in Western/European culture but by their relation to a specific determination/characterization of space. And this is exactly what has been basic and fundamental in the conception and/or thinking of the idea of culture in Latin America. For example, the famous and profusely commented classic seventeenth-century manuscript by Huaman Poma de Ayala (1534–1615) – *Nueva Corónica y Buen Gobierno* (*circa* 1615) – can indeed be interpreted in this context: it is the spatial position and condition that make 'Latin America' and 'Europe' equal/different, all the illustrations of the books, especially the 'maps', are an astonishing example of this *equal/reverse approach* (see for example Mignolo, 2005: 115 ff.). We will return to this issue in the last chapter.

4.3 From Imaginatio to Visuality

In *La relation critique* (1970) Jean Starobinski (1920–) has produced one of the most concrete, brief and accomplished histories of the evolution of the notion of *imagination* within Western/European culture. And what is interesting about this work is the fact that it allows us to observe a close connection between the definition of 'eye' and 'vision' with the idea of 'imagination'. If a conception of the function of the 'eye' cannot be separated from its public and social deployment as 'vision', then, equally, it cannot be detached from how it will function in respect to a *non-visible* domain. Paradoxically, within the Western/European culture, the visual refers not only to what is visible (sensible, perception) but also, and no less importantly, to what is not visible to the 'eye'. The most remarkable characteristic of Western/European imagination is perhaps this regular concern of the visual (visuality) with what is *not visible*. This situation had also been greatly present within the Latin American colonies, especially when it became entangled up with local *aborigen* concepts and traditions. Those concepts and traditions, which had been described (see for example Mignolo, 2005), were forced to be hidden, but never ceased to generate influences.

Starobinski's books show nicely how the Renaissance Western/European 'eye' is mainly concerned with art and historiography under two issues/problems: (i) one which refers to what is visible in the material sense and (ii) another which refers to what is not visible in the ordinary way. Quite rightly and appropriately Starobinski named this historiographical tradition 'L'Empire de l'imaginaire' ('The Empire of the Imaginary'). Thus, since the Renaissance period, the 'Western/European eye' has been concerned with two forms of 'vision': (i) the first of which referred to the public and material domain and (ii) the second of which referred to what is far away, reachable only by instruments or technical devices and/or by the re-creation of what is not there through images, that is, by the imagination. And what makes these two obvious forms of vision different is the fact that they constantly refer to each other: the visible is defined by what is not visible and vice versa. This situation within the Western/European domain was greatly developed thanks to the physical presence of the *colonies* and also to the idea of them.

Moreover, if we mention this evolution here it is because it was developed in Western/European environment at the same time as the first 'American colonies' began to be established, and is therefore part of the notion of 'colony' itself. Furthermore, these concepts of 'eye', 'image' and 'imagination' in their Western/European meaning were equally developed within the central and peripheral parts of the Empire(s). In fact, as we have shown elsewhere (see Canaparo, forthcoming), the imagination of space and the cartographic activities are fundamental to the settlement of the colonies in the so-called 'New World', at least in what we understand today as 'Latin America'.

Another aspect that Starobinski points out – even if it was not developed later – is the fact that, by situating Giordano Bruno (1548–1600) as the *figura/personaje agonal* (fighter, opposing personality, central protagonist/developer) in the evolution/creation of the Western/European meaning of imagination, we acknowledge one clear point: Bruno was the first Western philosopher in modern historiography to connect the notion of vision and imagination – as exchangeable terms – with the necessity of having a *cosmogony* or, to use a more contemporary term, to have a 'world-place': 'Les effets cosmiques, en nous, passent par la *vis imaginativa.*' (Satoribinski, 1970: 185). Without looking for it, Bruno established a perspective that Western/European cartographers and practitioner scientists would later develop: there is a close relationship between (i) our concept of vision, (ii) our idea of imagination and (iii) our perception of physical space. Moreover, those three aspects are to be understood and entangled in a single dimension that is going to be called – as indicated – 'cosmology' or 'world'. It is not a matter of chance that Bruno's works were developed at the same time as the first news about a 'new unknown world' was spread within the European territory and academies. Bruno was somehow *imagining* the spatial dimension of what was going to be known as 'new world' and it is exactly this imagination of 'worlds' or, more precisely, this imagination that is trying to construct worlds which, according to Starobinski, constitutes one of the most relevant and prolonged characteristics of Western/European culture. The close connection between thinking in philosophical terms and imagining what is not immediately visible was consolidated in the Western/European tradition around authors like Giordano Bruno: 'L'imagination est donc l'organe essentiel de notre relation au tout, le lien entre microcosme et macrocosme.' (Starobisnki, 1970: 185). The equivalent epistemological status that Bruno gave to what is physically visible and what is not immediately visible – but that can be 'visible' through *devices* – is indeed what cost Bruno his life, but more relevantly for us, is what makes him so contemporary to us, as Starobinski noticed. The assumption, for example, that what we understand scientifically as 'matter' – as well as the concepts deployed to talk about it – *is not visible* cannot be detached from this long Western/European tradition – equally, these conditions can be extended to art, politics, sociology, etc. As Starobinski also points out, Gaston Bachelard (1884–1972), among other Western/European authors, is someone who extensively explored this Renaissance *vis imaginativa* way, asserting that 'l'option scientifique et l'option imaginative sont à la fois polairement opposés, symétriques et complémentaires' (Starobinski, 1970: 192).

The construction of communities and countries in what we understand as Latin America cannot be detached from this tradition, less because it was imposed through the colonialism that designed 'America' than because the fact that the notion of colonialism itself – and all the concepts and terms attached to it – are grounded in this 'visual evolution'. Even more importantly, as we have already mentioned, this conceptual evolution was not developed 'only in Europe': without the colonies 'Europe' as a place could never be produced in its present dimensions, therefore many crucial cultural concepts like 'travel', 'translation', 'movement' and, of course, 'vision' and 'imagination', among others, were possible in the way we know them nowadays only because of the development they had within the colonies' territory and land or, at least, as a result of the interaction with the colonial space.

And thus some contemporary theories about colonialism – like the one that proposes a 'delinking' approach (see for example Mignolo, 2005) – are not naïve but are condemned to be without any real cultural power because they ignore the roots of the present concepts of Western/European culture that designed what is understood as 'Latin America' today. And, even acknowledging all the cruelties and the worst cultural consequences of Western/European colonialism/imperialism in Latin America, these *roots* of the conceptual system on the basis of which we produce meaning and talk about Latin American as such, cannot be ignored. The very interesting and useful book by Walter D. Mignolo on the 'idea of Latin America' (see Mignolo, 2005), despite all the strong arguments exhibited, is not very convincing in this respect: there is no *outside* to the Latin American space from which we can think and create concepts. And, paradoxically, it is the same W. Mignolo who, in *Local Histories/Global Designs* (2000), exposed one of the strongest reasons for this *culture limit* in Latin American thinking:

> The Occident, however, was never Europe's Other but the difference within sameness: Indias Occidentales (as you can see in the very name), was the extreme West, not its alterity. America, contrary to Asia and Africa, was included as part of Europe's extension and not as its difference. That is why, once more, without Occidentalism there is no Orientalism. (Mignolo, 2000: 58)

And, equally importantly:

> Occidentalism, in other words, is the overarching geopolitical imaginary of the modern/colonial world system, to which Orientalism was appended in its first radical transformation, when the center of the system moved from the Iberian Peninsula to the North Sea, between Holland and Britain. [...] Occidentalism – as I said – as the overarching imaginary of the modern world system was, at the same time, a powerful machine for subalternizing knowledge (from early

> missionaries of the Renaissance to the philosophers of the Enlightenment), and the setting up a planetary epistemological standard. (Mignolo, 2000: 59)

From this perspective there is not an *outside* because, even from the worst possible scenario, and as a distinction to what happens to Asian and African spaces and geographies, 'Latin America' was always *part* of the Western/European space – as a front line, as a limit, as a border line and as *frontera*, as previously indicated, but always *there*, as a peripheral part/dimension of the Western/European space. This situation allows us to partially understand why, within the 'Western imaginary', the shift of its centre from Europe to the US was accomplished in a very short period of time. And this also allows us to understand why the notion of centre/periphery was so important in many native theories and perspectives.

More generally speaking, this impossibility of establishing an *outside* of/for the local way of thinking is exactly the reason why many authors and artists were concerned – and still are – within Latin America with the *design and construction of the familial space*: not being able to be *outside* of the many colonial aspects of historiography and of the variety of elements of the present imperialism/financial globalism, then *images* – in the already quoted sense of *vis imaginativa* or *imaginatio* or *imaginario* – became an alternative not to being outside space but to generating spatial *itinerarios* ('itineraries') that can be used as an alternative interpretation to those generated by the above-mentioned historiography and the notion of 'present'. The development of an *imaginatio* allows us to establish connections and relations that sometimes writing – controlled by historiography – does not or, to be more exact, that writing is not yet able to put forward. And in this sense, for example, the *constructivismo* of Joaquín Torres García's (1874–1949) paintings, drawings and writings drives us into the notion of a *land without borders* present in Alejandro González Iñárritu's (1963–) works, constructing a possible line and evolution for this local *imaginatio* and by way of example.

Torres García's reversal map was an attempt to make a *local point* by acknowledging a condition of 'totality' or, to be more precise, the *local* was defined in Torres Garcia's map in relation to a conception of total space. The construction of a *local point* implied consideration of where the concept of space itself came from. González Iñárritu's short commercial film about BMW cars (2001) works in a similar way: the *local point* – a Latin American space, a Latin American conflict, a Latin American geography and urban space – is made out of a 'totality' condition (the universal condition of the car, the usual concepts of a war, the general idea of photography, the imperial army, Castilian song, the sense of displacement as a way to establish the notion of colony, etc.). However, the space captured by a cartographic perspective in Torres García's reversal map is approached by González Iñárritu in a more abstract way: speed, displacement and war are the real devices. If Torres García was dealing with imaginary colonial geographies, González Iñárritu is negotiating with an imperial sense of place where conceptualization is grounded in a general market condition.

The evolution from one 'cosmogony' that merges (i) the definition of 'eye', (ii) an idea of vision and (iii) a notion of imagination towards a conception and understanding of a world-place towards one 'history' and historiography of the imagination itself happens, as we have already suggested, within Western/European culture. It is something that can also be realized and tracked down within the idea and development of what is considered as *local* space in Latin America. And this is also why we said it is the concept of colonialism itself that has been developed *as much* in colonial lands and less, as had been postulated, in some 'central' part of the Empire (see for example Galeano, 1997). The works of Torres García, to keep to the example cited, functions exactly in this way: the construction of a 'universalism' grounded in local elements does not refer to common values but mainly to the necessity of formulating a conceptual *totalidad* ['totality']. At the same time, it is able to develop a condition which avoids any 'naturalism', that is to say, it is capable of constructing images in an 'artificial' sense that creates the local in an authentic and original form. And this construction of a local dimension from images is something that cinema and cinematography develop extensively (see Canaparo, forthcoming). The films of the commercially successful Mexican-born director González Iñárritu, as already indicated, can be considered as an example of this perspective which was already present in Torres García, among others.

5. 'El pensamiento de la cosa'

5.1 The Local as Ground

To *think spatially* also means exploring the entanglement already mentioned between having been/being at the borders of the Empire and trying to explore/explode the limits of Western philosophical tradition – even the authors that presented themselves as the most critical to Western/European philosophies could not avoid commenting, using, discussing and/or referring to the Western/European philosophers/philosophies (see for example Mignolo, 2000: Chapter 1). The notion of *geo-epistemology* must also cover this activity and/or attempt.

What has been criticized as colonial cultural behaviour, that is, the passive use of the European cultural tradition as a way of constructing a sort of 'Latin American identity', has been established today as a critical diagnosis and does not need further consideration (see for example Mignolo, 2005). However, what merits closer and more detailed thought is the fact that, contrary to what has been suggested, a radical 'delinking' attitude towards past and present cultural production in Europe is not an intellectually available possibility (on the subject see for example Mignolo, 2005; and in particular Amin, 1990). Simple observation of the university system and the cultural and intellectual publications within Latin America should be enough to show that the millions of people who inhabit those areas and institutions are not going to dismantle their values ('system of belief'), points of reference ('historiographical system') and lexicon ('language domain') – three aspects that we can consider from now on as the *ground*. Furthermore, the fact that the consideration of culture and thinking is something that happens *almost exclusively* in academia and/or universities, or in publications or institutions connected with them, as is also the traditional and historical case in Western/European culture, is another clear situation that makes any 'delinking' perspective illusory. The original work of Samir Amin (1931–) regarding this subject was published in 1985 and was entitled *La déconnexion* (see Amin, 1990). The fact that this original sense of 'delinking' was designed to address economic issues within a neo-Marxist context and designed for a world conceived without the radical technological changes that have happened recently, makes the suggestion of reformulating the notion of 'delinking' even worse.

What indeed might be feasible, as Mignolo (2005) also considered, is the possibility of expanding this *ground* by incorporating more radical and innovative approaches to the ways in which we can make use of European products and traditions. More practical philosophical cynicism and less concern about the moral or political considerations pertaining to historiographical narrations should be the *motto*. Within a perspective of *spatial thinking* we are free from the need to produce any 'independence' or 'liberation' from Western/European cultural objects and ideas because what is relevant is elsewhere. However, in line with this dominant and past perspective of considering the Western/European culture as a *foreign element*, the local translation and use of those Western/European cultural objects and ideas has been understood by some authors as a prolongation of colonialism and/or 'subaltern' forms of knowledge. The *reversal thinking* we are proposing here should bring simplicity and operability to this approach: *thinking spatially* is what makes the difference. The foreign aspects of Western/European tradition – if there are any – do not contribute to a better understanding and/or operability towards the emergence of a form of local knowledge.

Constructing practical principles to operate at the borders of the Empire and at the borders of that Empire's culture, knowing that one *is there*, is an indispensable strategy for survival. Furthermore, we should consider, as many authors do nowadays, that the 'borders' act less as a limit and more as a *frontera* ('borderline'). In this way we do not need to go to the physical limits of a territory and/or far away from the capitals – as local people believed in the nineteenth century (see for example Canaparo, forthcoming) – to find these borders, sometimes quite the contrary. Thus, the adoption of these principles works towards a cynical, philosophical, non-European approach, and seems to us to present a positive intellectual and cultural opportunity. In our view, this perspective goes back to the first attempts of Huaman Poma and Inca Garcilaso (1539–1616), to give two historiographical examples mentioned previously. Cynicism as an intellectual perspective is at the border of what has been understood as Western philosophy (see for example Sloterdijk, 1993). The most persistent phenomena of the cynical perspective is a paradoxical attachment to the environment characterized by an extreme de-localization. Or, to put it in more simple terms, it is a paradoxical way of thinking that, by radically defining the sense of what is local, achieves a punctual and empirical meaning of what is understood as immediate environment. It is in this sense then that we postulate a cynical perspective.

And under these conditions, and in a historiographical context, not only are Human Poma and Garcilaso examples, but also characters like Lautaro (*circa* 1534–1557), who was 'educated' when he was a slave, or Manuel Baigorria (1809–1875), the so-called 'cacique blanco', who was always confusing military preferences and ethnic belonging, or the Indian Mariano Bejarano (1837–1911) who became a soldier in order to survive, or the political boss ('cacique') of Ranquel, Baigorrita, whose early colonial education lead him to an extreme sense of defeat and who, when captured demanded to be killed. These are some random examples from the Southern Cone. However, the notion of 'daño' or 'herida' as a basic and fundamental form of education is at the centre of the colonial experience – and the cynical perspective also tries to acknowledge it. The comments of Mignolo about the 'colonial wound' makes, I believe, the same point (see Mignolo, 2000 and 2005).

An intellectual, cynical perspective refers to the possibility of being able to work *at the same time and at the same level* with radically different traditions and perspectives that are not complementary to one another but rather in *conflict*. This perspective refers to individuals who, for one reason or another, have been dealing with at least two different systems of belief, two different sets of references, and two different language domains.[14] It is this approach, then, that we define as philosophical cynicism, and which has been addressed by other authors as 'an other thinking' (Mignolo, 2000: 66 and ff.) or 'double critique' (Mignolo, 2000: 68). We will return to the subject of a local cynical philosophical tradition in Latin America when we discuss the notion of *reversal thinking*.

14 Authors like Martin Lienhard describe and analyze this duality in terms of 'cultural diglossia' (see Lienhard, 1992).

What is relevant in all this is not only the fact that we might be able to consider European cultural production in another light. It is also fundamental that we might be able to produce a new way of generating concepts by constantly switching our point of view and, contrary to what has been happening since the nineteenth century, abandoning any idea that one perspective should prevail over all the other possibilities. For example, we should avoid the definition of interpretation – or the heuristic attitude – that looks for one meaning according to the content of a particular text, as many of the literary critics and professors have been doing/teaching locally for the last 100 years. Furthermore, the paratextual aspect of a written work should be as important as what we still call 'content'. The 'peripheral' context of reception and the 'peripheral' context of production of that written text – to keep the example – should also be as important as the 'content'. The market consideration of this written work and its existence as a material object should also be fundamental aspects to consider in the production of meaning. Because this is precisely the point: the idea of interpretation as a sort of academic tool or literary concept should be abandoned in favour of a wider cultural perspective where what is at stake is the production of meaning in a global system and, still more relevantly, the location of our space of reference within that global system.

However, we need to acknowledge that this proposal creates a paradox because if, on one hand, we pay more attention to a strategy that considers the existence of a global system, on the other hand, it is necessary to develop local concepts or, more specifically, to have a local axial space – what we will call 'epistemological ground' – upon which we base all conceptualization or, what is equal within this context, upon which to determine a 'loci of enunciation'. And this is the same paradox that, from other perspectives, has also been indicated by authors like W. Mignolo using the opposition of 'local histories' vs 'global designs' (see for example Mignolo, 2000).

And this construction that we are proposing here drives us in the direction of local/locality or, more precisely, to how we will define the local when it is considered as the necessary ground for living. And thus we arrive at the consideration of the 'local' as ground in three forms that are entangled (see Table 1.2).

Ground 'stricto sensu'	Belief system	Historiographical system	Language domain
Epistemological ground	Defined as *axial space*		
Cultural Ground	Biological dimension	Body dimension	Object dimension

TABLE 1.2. The three meanings of 'ground' as considered in relation to the perspective of *spatial thinking* and in relation to the interaction between 'philosophy' and 'geography' as discussed above.

The ground that we can indicate as 'stricto sensu' refers to three aspects. Firstly, to a system of beliefs in the empirical sense, that is, to the system upon which individual and corporate decisions relies publicly. Secondly, to a historiographical system that not only refers to publications or books but also to the *imaginary* created from an established tradition of knowledge. And, thirdly, to a language domain that not only refers to an 'idiom' but that is also considered as *storage* and memory (see Table 1.2).

The ground indicated as 'epistemological' is the collected form of ground, is the axial space where 'geography' and 'philosophy', as described before, become entangled. This ground is also a *connector* between the traditional ground (the ground 'propiamente dicho', *stricto sensu*) as described, and the 'cultural' ground.

The ground indicated as 'cultural' is a physical dimension and refers to three elements that work together. The first is a biological dimension that functions as both a deterministic and a random element at the same time. The second is a bodily dimension that works as imaginary and as a concept. And, finally, the third is a notion of the object as holistic taxonomy and as a world-system. The fusion of these three dimensions is what we will call 'living space', assuming that any definition of local 'culture' only can be biologically orientated.[15]

15 For a discussion on this issue see for example Maturana, 1995a and 1995b.

The result of placing together 'geography' and 'philosophy' in relation to Latin America offers a paradox effect: 'geography' becomes something non-universal and locally situated and 'philosophy' becomes thinking (see Table 1.4).[16] Yet if we accept the significant role of space within Latin America, then the notion of *cosa*, as discussed here, becomes the axial element that allows us to connect the local dimension (the non-geography) and the thinking (the non-philosophy).[17] Expressed in local terms, the *axial space* that we have just situated in an *agonal* (from 'agora' and 'dromos') position, will be understood, under the conditions of the thinking previously indicated, as *cosa* ['thing']. The *cosa* will be the concept of concepts by contraposition to the Cartesian 'I' or to the 'philosophy of being' exposed by Heidegger and which can be considered as the most accurate version of Western/European philosophy during the twentieth century (see for example Mignolo, 2005 and Edwards, 2007). The *cosa* will be the link between 'geography' and 'philosophy'. And that is when the notion of 'universalism' attached to the *cosa* becomes conflictual. The question of *universalism*, as we indicated already, drives us, within the Latin America environment, into the dimension of the *cosa*, understood as basic *ground* (see Table 1.2). And one of the aspects that characterizes Latin America as a peripheral area in this context is in the fact that the European *universalism* introduced by the notion of *cosa* – from Kant to Heidegger – functions locally *in absentia*. In this sense it is not the 'cosa' but *its absence* that counts and this is also why the notion of ground always presents difficulties in Latin America in relation to derived terms like 'land', 'territory', etc. We can also express this in Heidegger's terms: we have the 'cosa' but without the *cosidad* (*choséité*) (see Heidegger, 1971: Part A). And this situation also partially explains the local popularity and success of theories that somehow appeal to the *cosa* as the dominant characteristic. Ricoeur's (1913–2005) notion of the *chose du text* could

16 Even Heidegger, when discussing the notion of *cosa*, refers it to questions of distance, belonging, object, speed and far away (see Heidegger, 1971: Part A).

17 Regarding the notion of 'cosa' as an axial space – in the physical and conceptual sense – see also Canaparo, 2007.

be considered as a paradigmatic example here (see Ricoeur, 1986: 125 ff.).

There is not then *one* definition of *cosa* but many. *Cosa* is a result of a particular construction and the fact that we always name 'cosa' with different set of meanings and circumstances is a result of (i) the necessary *continuity* ('durée' in terms of Bergson) that requires the idea that we always live in a present time and is also due to the fact that (ii) a sort of 'principle of reality' is necessary as well as (iii) coordinates ('coordinadas') or *points de repère* to construct a 'world-idea'. These three regular aspects of 'cosa' are what support the notion of ground in general terms and, ultimately, the idea of space itself.[18]

5.2 'La cosa de las cosas'

Let us consider an example. A recent new edition of a book by the already quoted Ezequiel Martínez Estrada has generated a sort of polemic within the River Plate environment regarding the 'content' of the book ('los dichos-escritos del libro'), but it is also interesting due to the context in which the author originally situated the book. Published in its first edition in 1956, *¿Qué es esto?* was presented at the time by Martínez Estrada as 'catilinaria'. The 'catilinaria' refers, as is well known, to the four letters written by Marcus Tullius Cicero (106–43 BC) in 63 BC against Lucius Sergius Catilina (*circa* 108–62 BC) who was then attempting to overthrow the government of Rome. What is interesting here is the fact that Catilina made his name and career in Africa, which is at the 'border' of the Roman Empire, a situation that allowed him to be presented as a Consul candidate from *outside* Rome. The fact that the 'catilinarias' were published by Martínez Estrada after Juan Domingo Perón's (1895–1974) defeat makes the immediate and direct application of the plot to the political local situation all the more ambiguous.

In fact, a book like *¿Qué es esto?*, if considered from the European perspective of modernity, will be judged by its *political commentaries*.

18 Regarding this relationship between ground, space, and conceptualization, see Plata/Canaparo, forthcoming.

However, if analyzed from a more local, spatial perspective, it will overcome that single possibility and tackle, for example, the notion of 'esto' as *cosa* ['thing'], thereby being a way of understanding the conceptual evolution and conditions of production of meaning in which Martínez Estrada conceived its work. The *geographical pertinence/ belonging* to a place or land or system of objects is always established in peripheral areas under conditions of questioning and/or threat. Moreover, the same idea of *geography* as an area/space that indicates 'being part of' is not adequate to address the situation of Latin America as *a particular* perspective or understanding.

The traditional Western/European conception of geography as the simple study/approach of localities has been questioned by several authors, such as European anthropologists and urbanists themselves, who discus the traditional notion of geography as a nineteenth-century conception and perspective (see for example Capel, 2001). What seems to be obvious is that the implementation of *naturalism* from an aesthetic point of view, and the implementation of an idea of *nature* and of an attached *natural philosophy* as a World-idea, cannot be separated from two regular conditions. These are (i) the necessary monopoly of *one single view* about 'nature' and 'reality' that this perspective deployed, and (ii) the impossibility of thinking or conceiving meanings for concepts that go outside that World-idea.

¿Qué es esto? should be a tool – as we will see – for exploring concepts like the idea of 'esto' as *cosa* or the notion of 'geography' and not an object which adds yet another element of dissent within the colonial space itself. In this case – as in many others – the possible local distinctions that this book can generate are less important than how they can be situated in the context of a peripheral space: Why did Martínez Estrada use and deploy metaphors as a methodology of analysis? Why are all of them from contemporary European history? Why aren't there *expectations* within the arguments? Why is the idea of 'Perón' as a *personaje público* ['public personage'] considered more relevant than as a narrative character?, etc.

What we are suggesting here is that, in constructing a local space, the elaboration of concepts became the relevant dimension/activity, not the interpretation of them or the way in which they have been presented in the media. Of course, the idea is not to question the entire

Western/European conceptual evolution using a book by a River Plate writer published in 1956. However, a change in the sense in which historiography and written activities have been traditionally considered, following more from Western/European perspectives and not so much the local particularities, is indeed a point of departure. We will call this perspective *reversal thinking*.

The ability to situate books like *¿Qué es esto?* under another historiographical category – and under another view of writing – than the one referred to in its content is what we are talking about here. The ability to completely and radically change ideas and perspectives about a book without changing the book itself is part of what we call *reversal thinking*. The ability to radically change meanings with regard to a book without changing it as 'object', by modifying it as *cosa* is another aspect of this kind of thinking. How to modify the epistemological conditions of meaning of a particular *material culture* without refusing this *material culture* as point of departure is the challenge here. And this is also why we believe that such a perspective should include the ability to confront a public domain populated by objects and/or things designed and created by Western/European culture or by a perspective grounded in it. Within the present public domain and market, and within the dominating historiography, the Western/European perspective controls massively the meanings and the semantic dimension of Latin American space – to try to ignore it, or to try to 'delink' it, is a quick way to further help the consolidation of universities and publishers that are undoubtedly Western/European-orientated and, thus perpetuate the 'Western imagination' as defined by Mignolo himself (see for example Mignolo, 2000).

Moreover, the possibility of radically changing the already established historiographical readings and statements about this book by Martínez Estrada is what this new, flexible perspective of *reversal thinking* should allow us to do. An historiographical tradition – fundamental for a culture that situates the written dimension at the core of its intellectual activities (see for example Canaparo, 2005) – should be a tool for the present and not a phenomenon of/for the past. Being able to modify our ideas/conceptions of the past and our perceptions of the future is one of the consequences of this proposed change.[19] We should learn about this from the Western/Europeans authors instead of being concerned about how bad their colonialism/imperialism was/is or instead of trying to reproduce exactly what they do. We are not the same and we never will be: this is not a political or academic decision, it is simply a physical and mental fact. And as we have indicated already the notion/perspective of knowledge and the situation/location of it are the two aspects that lead us most quickly and directly into this question of the *difference*.

But *¿Qué es esto?* is more than just the 'catilinaria' that its author considered it to be. The book also displayed an interesting local aspect that somehow could be considered as a regular 'principle' within Latin American space. When we want to ask a fundamental 'philosophical' question, when we want to ask 'la pregunta primera' in the Aristotelian sense, then we use the notion of *esto* as 'cosa' in the sense of basic ground. So, more than for its surrounding political circumstances, this book by Martínez Estrada is relevant for the very question deployed to raise the problem: the question of *esto* is the question about the nature of a problem and a definition of the problem itself. The question of *esto* is the question about culture in the sense of *material culture* stated before, that is, about the epistemological conditions in which a particular notion of culture is developed and used.[20] The question of *esto* is the question about la 'cosa' in the speculative terms of local thinking (see Table 1.4). And it is also for this reason that we believe *thinking spatially* is a way to understand the peripheral condition of knowledge within Latin America.

Thus, *thinking spatially* could be considered as equivalent to 'el pensamiento de la cosa' ['the philosophy of the thing'] and could be addressed in three different ways: (i) as 'el concepto de cosa' [the concept of thing], (ii) as 'la cosa del concepto' (the thing of the concept), and (iii) as 'la cosa del pensamiento' [the thing of thinking] (see Table

19 We will return to this subject later on Chapter 4.

20 The work of the British academic Daniel Miller, *Material Culture and Mass Consumption* (1991), has some aspects similar to the definition postulated here.

'Cosa'	'El concepto de cosa'	'La cosa del concepto'	'La cosa del pensamiento'
Modes of 'cosa'	'Cosa' as something ('algo', *aliquid*)	'Cosa' as immediacy unit (*unum*)	'Cosa' as truthfulness (*verum*) and goodness (*bonum*)
'Spatial Thinking' domain	Ground Language Understanding	'Mundo' Local environment	Present Becoming

TABLE 1.3. The domains of the *cosa* ['thing'] that function as elemental characteristics of 'el pensamiento de la cosa' and as a natural deployment of the perspective about space.

1.3). The first approach refers to the dimensions of *lexico* and *ground*. The second approach refers to the dimension of *mundo*. And, finally, the third approach refers to the dimension of 'devenir' (becoming), 'acontecer' (*durée*, event) and 'movimiento' (movement).

In this way Martínez Estrada's 'esto' can be considered as a local *cosa* in the three senses: (i) as a basic meaning about the local language and understanding ('cosa' considered as something), (ii) as a conception about the local environment ('cosa' considered as 'immediacy unit'), and (iii) as a *point de repère* of the present ('pure becoming') ('cosa' considered as the truthfulness and goodness). Stating that *esto es una cosa* ('this is a thing'), in the context of our example, is a way to get to the basic meaning and to the fundamental question about locality and about what is considered as characteristic of the local space.

In the context of the proposed *spatial thinking*, as an alternative to the monopolist Western/European philosophical historicism, that considers temporalization as the *driving element* of modern reason/rationality, the notion of *cosa*, instead of the concept of 'being' or 'verb' – like basic forms of the Cartesian 'I' (see for example Mignolo, 2000: 91–126; and also Mignolo, 2005: 57–72 and 135–141) – is what is proposed here in the context of the example mentioned. Of course we are not referring here to the scholastic and 'transcendental' notion of *cosa* but to a sort of material and imaginary world-view ('res in se [vel per se] spectata') understood from the three dimensions indicated in Table 1.3. The *cosa* is neither the *Ding an sich* from Kant nor the expression of any ultimate scientific reality.

5.3 The Space of Thinking

Within *el pensamiento de la cosa* ('thinking the thing', 'thinking of things'), derived from the *spatial thinking*, that is, from the assumption of the fact that any thought is a 'spatial thought' and that any language is a 'geographical idiom', we are pushed in the direction of the 'borders' or, as Mignolo already indicated, to 'border thinking' (see Mignolo, 2000: 49–90). And at this point it is also useful to return to a conceptual distinction initially made by Friedrich Nietzsche and re-invented later by many European authors, such as Martin Heidegger, for example. I am referring to the distinction made between 'thinking' and 'philosophy'. These are two activities that, according to Nietzsche, should be considered as radically different. According to Heidegger, who follows on from Nietzsche's *dictum*, 'thinking' is a poetic and artistic activity that also somehow connects with a 'primitive' state of mind. On the contrary, 'philosophy' is a product of civilization and has a historical purpose. It also has a disciplinary dimension: philosophy in Western/European tradition has always been related to academies. Given this context, the temptation to consider Western/European 'thinking' as *closer* than 'philosophy' to the colonial environment has always been there and a number of authors have made use of this possibility. The idea, for example, that only the *metaphysical* aspects of the Western/European theories can offer some degree of freedom to non-Western/European thinkers is not new and has been present in Latin America since the existence of the avant-garde aesthetics movement at the beginning of the twentieth century.

To some extent, *spatial thinking* will undoubtedly find more elements to construct a *familiar environment*, in a given present, within a visual perspective rather than within a more authoritative and institutionally developed theory of writing, even against the view of a great number of local writers and analysts. Furthermore, from the conceptual point of view, 'philosophy' is far more formalized than 'thinking', a definition which always exhibits conflicts and disagreements. Culture,

'Philosophy'	Context of Civilization	Theory of Writing	Domain of Geometry
'Thinking'	Limits and borders of Culture	Visual Perspective	Domain of *Imaginatio*

TABLE 1.4. The two domains of the *naturaleza especulativa* ['speculative nature'] within Western/European culture.

understood as *geography* in the Western/European context (see Table 1.1), has normally acted as a connection between these two different aspects of 'philosophy' and 'thinking'. While 'philosophy' gives to culture the ultimate conceptual context (the *territory* in ground terms), 'thinking' always looks to the limitations and unsignified meanings of it (the *land/new land* in ground terms). We understand better here why Heidegger himself uses spatial terms like *Grundbegriffe* or 'concepts of ground' to discuss the subject (for a more in depth discussion on the matter see Heidegger, 1967, 1968 and 1993). And it is based on this 'culture' situation that authors working in colonial environments find in 'philosophy' a sort of continuation of colonialism or, to be more precise, a development of 'colonial historicism'. While the same authors find in 'thinking' a sort of *intersticio* (see for example Nash, 2001 and Díaz Gajardo, 2002), a sort of exhibited crisis where it should be possible *to see the real conceptual scheme* of Western/European culture. The autobiographical aspects of the works of Rodolfo Kusch (1922–1979) seem a very pertinent example here (see for example Mignolo, 2000: 149 ff., and 304 ff.).

If 'philosophy' constitutes the domain of what is known, 'thinking' stands for what is present and uncertain, in which case, if 'philosophy' allows authors and critics to situate themselves with regard to Western/European culture in terms of *similitudes*, 'thinking', on the contrary, forces authors and critics to look at what is *different*, to what is not there, to what is non yet conceptualized and part of the 'pure becoming' by which we understand our perceptions. 'Philosophy' shows the way of society through historiography, while 'thinking' shows the way of what is *not social* through art and poetry. In this context, we can understand why many of the so-called Latin American *native* philosophies are presented, in relation to the dominant Western/European culture, as strategies for survival (see for example Galeano, 1997). If the extremely local version of the Western/European philosophical imaginary pushes authors in the direction of a theory of writing, following thinkers like Heidegger, Derrida and Deleuze among others, then the *reverse* of this 'philosophy' is, in the context of our discussion here, a non-temporal, visual perspective that can be indicated locally as 'thinking'. And this explains why, among other reasons, Latin America produces more *ensayistas* ('non-fiction authors') than philosophers and more fiction writers than any other colonial space on the planet.[21] 'Thinking' then is not a natural or telluric fatality – as local authors like Martínez Estrada (1889–1962) have postulated (see for example Martínez Estrada, 1940 and 1993) – but a practical result of the conditions of a peripheral space.

In Table 1.4 the three main characteristics that classically differentiate 'philosophy' from 'thinking' are presented. And it is under these characteristics and distinction that we will postulate in Chapter 3 the relevance and significance of a theory of translation in which (i) civilization needs to be understood as a 'border culture', (ii) theory of writing as a 'visual culture' and (iii) geometry as 'imaginatio'. When we said that what in Western/European philosophical imaginary is *time*, becomes in Latin America *space*, this was the *translation* – or theory of translation – that we were referring to.

If being able to 'think' in conceptual terms within a colonial environment means that we need first to be aware of the limits and borders of Western/European culture, because this culture is part of the colonial space itself and we are not able to construct an *outside* to it, then, as we have already shown, a notion of *spatial thinking* is not only a possible alternative to the 'colonial historicism' that dominates the Western/European historiography, it is also a way of being able to create and/or re-construct a notion of space in relation to the meaning of the

21 In 1976 the Frankfurt Book Fair, the most important book fair in the world in market terms, was dedicated to Latin America and one of the slogans exhibited there was 'América Latina: el continente literario por excelencia' ('Latin America, the literary continent by excellence'), following this very perception of the colonial literary space *by default*.

concept itself. This point is fundamental, in our view, to conceiving a real functional alternative to Western/European control of the academic and editorial market.

With Figure 1.1 in mind, it is also now important to say that the way in which colonialism was established was grounded on strong *enlightenment principles*. And, paradoxically, many of the most reputed anti-colonial theories about Latin America during the twentieth century – as we will see later – were also thoroughly grounded on 'ideas iluministas' [enlightened ideas]. The pedagogical theories of Paulo Freire (1921–1997) seem an astonishing example to me (see for example Freire, 2000).

However, less important than determining whether or not we think in terms of enlightenment principles is the acknowledgment of the fact that thinking was constructed in a Western/European sense and, equally, in a colonial sense, from a perspective of historiography and politics that has always associated conceptual speaking 'knowledge', 'education' and 'learning' as a part of a social/intellectual process (referred to before as 'comprehension') which aimed at determining the meaning of a particular culture (see Figure 1.1). And if this process, as indicated before, exposes/shows the limits of the Western/European perspective it is because we are already at the limit of what is *conceptually possible*. It is not that many European philosophies within the twentieth century were focused on this problem where meaning ('semantic' dimension) and event ('pure becoming' dimension) conflict with each other. From Heidegger to Derrida, despite other academic subjects, this issue has always been considered as a priority for Western/European philosophers – a situation which also explains why European authors with a very different approach, for example D. Derrida and P. Bourdieu (1930–2002), referred to Heidegger's work as a way of establishing their respective *ground* in their fields of work.

Finally, this proposed *spatial thinking* is also a way of acknowledging one of the most important regular characteristics of the Latin American space. From the *Comentarios Reales de los Incas* (1609) by Inca Garcilaso de la Vega to *Postmodernidad y transmodernidad. Diálogos con la filosofía de Gianni Vattimo* (1999) by Enrique Dussel, we believe there is always an equivalent spatial principle in the way the theoretical systems and schemes from local/colonial authors are constructed and established. The *conceptual discovery* of Latin America is something that paradoxically happened in 'Europe' or the 'United States', that is, in the central part of the Empire. What's more, the spatial construction of Latin America happens as a reaction to the perception/acknowledgement of the Western/European territories. And this also why, as we will see later, the theories that postulate an *outside* perspective of the space of colonialism are condemned to work in an artificial atmosphere or 'espacio vacío' environment [empty space, 'zero atmosphere']. W. Mignolo's proposal, which tries to associate the developments and products from or about the 'chicano culture', within the US, with the whole meaning and space of Latin America is, we believe, a very good example of this spatial emptiness (see Mignolo, 2005).

6. Bibliography

Amin, Samir. *Delinking. Towards a Polycentric World* (London: Zed Books, 1990).
Apel, Karl-Otto. *Charles S. Peirce: From Pragmatism to Pragmaticism* (Amherst: University of Massachusetts Press, 1981).
———. *Understanding and Explanation: A Trascendental-Pragmatic Perspective* (Cambridge, Mass.: MIT Press, 1984).
———. *Le logos prope au langage humain* (Paris: Editions de l'éclat, 1994).
———. *Semiótica transcendental y filosofía primera* (Madrid: Sintesis, 2002).
———. *L'*a priori *du corps dans le problème de la connaissance* (Paris: Cerf, 2005).
Badiou, Alan. *Being and Event* (London: Continuum, 2005).
Bartky, Sandra Lee. 'Heidegger and the Modes of World-Disclosure', in *Philosophy and Phenomenological Research*, Vol. 40, No. 2 (Dec., 1979), pp. 212–236.
Bezza, Giuseppe. 'What is classical astrology?', in *Schema*, 1, June (1986), pp. 1–4.
Bhabha, Homi K. *The Location of Culture* (London: Routledge, 1994).
Blumenberg, Hans. *La legibilidad del mundo* (Barcelona: Paidós, 2000).
Canaparo, Claudio. *Imaginación, mapas, escritura* (Buenos Aires: Zibaldone, 2000).
———. *Ciencia y escritura* (Buenos Aires: Zibaldone, 2003).
———. 'La consumación del realismo', in G. Fabry/C. Canaparo, eds., *El enigma de lo real. Las fronteras del realismo en la narrativa del siglo XX* (Bern: Peter Lang, 2007), pp. 199–275.
———. *El imaginario Patagonia. Ensayo acerca de la evolución conceptual del espacio* (forthcoming).

Capel, Horacio. *Dibujar el mundo. Borges, la ciudad y la geografía del siglo XXI* (Barcelona: Ediciones del Serbal, 2001).

Collins, Randall. *The Sociology of Philosophies. A Global Theory of Intellectual Change* (Cambridge, Mass.: Harvard University Press, 1998).

Deleuze, Gilles/Guattari, Felix. *Capitalisme et schizophrénie 2. Mille Plateaux* (Paris: Minuit, 1980).

———. *Qu'est-ce que la Philosophie?* (Paris: Minuit, 1991).

Derrida, Jacques. *Marges de la philosophie* (Paris: Minuit, 1972).

———. *Of Grammatology* (Baltimore: Johns Hopkins University Press, 1976).

———. *Writing and Difference* (London: Routledge, 1978).

———. *Edmund Husserl's Origin of Geometry: An Introduction* (Lincoln: University of Nebraska Press, 1989).

———. *Le problème de la génèse dans la philosophie de Husserl* (Paris: PUF, 1990).

———. *Mal d'Archive* (Paris: Galilée, 1995).

———. *The Problem of Genesis in Husserl's Philosophy* (Chicago: Chicago University Press, 2003).

Díaz Gajardo, Víctor. 'Fragmentación cultural y memoria histórica', in *Ciudad arqueológica. Ciudad Virtual de Antropología y Arqueología*, Congreso 2002, 8 pp.

Dussel, Enrique. *The Invention of Americas* (New York: Continuum, 1995).

Eco, Umberto. *Semiotics and the Philosophy of Language* (Bloomington: Indiana University Press, 1986).

Edwards, James. 'The Thinging of the Thing: The Ethic of Conditionality in Heidegger's Later Work', in H. L. Dreyfus/M. A. Wrathall, eds., *A Companion to Heidegger* (London: Blackwell, 2007).

Enaudeau, Corinne. *La paradoja de la representación* (Barcelona: Paidós, 1999).

Faye, Jean-Pierre. *Qu'est-ce que la philosophie?* (Paris: Armand Colin, 1997).

Feyerabend, Paul. *Problems of Empiricism* (Cambridge: Cambridge University Press, 1981).

———. *¿Por qué no Platón?* (Madrid: Tecnos, 1985).

Foerster, Heinz von. *Observing Systems* (Seaside, Ca.: Intersystems Publications, 1981).

———. *Understanding Understanding* (New York: Springer, 2003).

Foucault, Michel. *The History of Sexuality 3. The Care of the Self* (London: Penguin, 1990).

———. *The Hermeneutics of the Subject* (New York: Pagrave/Macmillan, 2005).

Freire, Paulo. *Pedagogy of the Oppressed* (London: Continuum, 2000).

Galeano, Eduardo. *Memory of Fire 2* (London: Quartet, 1986).

———. *Memory of Fire 1* (London: Meuthen, 1987). [1987a]

———. *Memory of Fire 3* (London: Quartet, 1987). [1987b]

———. *Open Veins of Latin America: Five Centuries of the Pillage of a Continent* (New York: Monthly Review Press, 1997).

Galison, P./Thompson, E., eds. *The Architecture of Science* (Cambridge, Mass.: MIT Press, 1999).

García Canclini, Néstor. *Culturas híbridas. Estrategias para salir y entrar de la modernidad* (Buenos Aires: Paidós, 1992).
Genette, Gérard. *Paratexts: Thresholds of Interpretation* (Cambridge: Cambridge University Press, 1997).
Glasersfeld, Ernst von. *The Construction of Knowledge – Contributions to Conceptual Semantics* (Salinas, Ca.: Intersystems Publications, 1987).
———. *Radical Constructivism. A Way of Knowing and Learning* (London: Falmer Press, 1995).
Groussac, Paul. *El viaje intelectual: impresiones de naturaleza y arte* (Buenos Aires: Librería General de Victoriano Suárez, 1904).
Gutiérrez, Gustavo. *A Theology of Liberation: History, Politics and Salvation* (Maryknoll: Orbis Books, 1988).
Heidegger, Martin. *What Is Philosophy?* (New Haven: Yale University Press, 1958).
———. *What Is a Thing?* (Chicago: Henry Regnery, 1967).
———. *What Is Called Thinking*? (New York: Harper and Row, 1968).
———. *Qu'est-ce qu'une chose?* (Paris; Gallimard, 1971).
———. *Basic Concepts* (Bloomington: Indiana University Press, 1993).
———. *La cosa* (Barcelona: Ediciones del Serbal, 1994).
Husserl, Edmund. *L'origine de la géometrie* (Paris: PUF, 1962).
Lakoff, George/Johnson, Mark. *Philosophy in the Flesh: The Embodied Mind and its Challenge to Western Thought* (New York: Basic Books, 1999).
Lampert, Laurence. 'On Heidegger and Historicism', in *Philosophy and Phenomenological Research*, Vol. 34, No. 4 (Jun., 1974), pp. 586–590.
Latour, B./Woolgar, S. *Laboratory Life* (Milton Keynes: Open University, 1979).
Lebeau, André. *L'engrenage de la technique* (Paris: Gallimard, 2005).
Lévy, Jacques. *Le tournant géographique. Penser l'espace pour lire le monde* (Paris: Belin, 1999).
Lienhard, Martin. *La voz y su huella* (Lima: Horizonte, 1992).
Lyotard, Jean-François. *The Postmodern Condition* (Manchester: Manchester University Press, 1984).
———. *The Differend: Phrases in Dispute* (Minneapolis: Minnesota University Press, 1988).
Martín-Barbero, Jesús. *Al sur de la modernidad* (Pittsburgh: IILI, 2001).
Martínez Estrada, Ezequiel. *La cabeza de Goliath* (Buenos Aires: Club del Libro, 1940).
———. *Radiografía de la pampa* (Buenos Aires: Fondo de Cultura Económica, 1993).
———. *¿Qué es esto? Catilinaria* (Buenos Aires: Biblioteca Nacional, 2005).
Maturana, Humberto. *La realidad: ¿objetiva o construida? I. Fundamentos biológicos la realidad* (Barcelona: Anthropos, 1995). [1995a]
———. *La realidad: ¿objetiva o construida? II. Fundamentos biológicos del conocimiento* (Barcelona: Anthropos, 1995). [1995b]
Michaud, Yves, ed. *Les États de la matière* (Paris: Odile Jacob, 2002).

Mignolo, Walter. *The Darker Side of the Renaissance. Literacy, Territoriality and Colonization* (Ann Arbor: Michigan University Press, 1995).
———. 'Espacios geográficos y localizaciones epistemológicas: la ratio entre localización geográfica y la subalternización de conocimientos', en *Pensar* [Bogotá], número 34 (1996), available at www.javierana.edu.co/pensar/Rev34.
———. *Local Histories/Global Designs. Coloniality, Subaltern Knowledge and Border Thinking* (Princeton: Princeton University Press, 2000).
———. *The Idea of Latin America* (London: Barckwell, 2005).
Miller, Daniel. *Material Culture and Mass Consumption* (London: Blackwell, 1991).
Monsiváis, Carlos. *Los rituales del caos* (Mexico: PFC/Era, 1995).
———. *Aires de familia. Cultura y sociedad en América Latina* (Barcelona: Anagrama, 2000).
Montgomery, Scott L. *The Moon and the Western Imagination* (Tucson: University of Arizona Press, 1999).
Nash, Mary. 'Intersticios, zonas de contacto intercultural y la construcción de nuevas identidades urbanas', in www.ub.es/mutingen/, 2001, 6 pp.
Nietzsche, Friedrich. *The Gay Science* (New York: Random House, 1974).
———. *Philosophy in the Tragic Age of the Greeks* (Chicago: Random Regnery, 1962).
Nisbett, Richard E. *The Geography of Thought. How Asians and Westerness Think Differently... and Why* (New York: Free Press, 2003).
Pais, Abraham. *'Subtle is the Lord...'. The Science and the Life of Albert Einstein* (Oxford: Clarendon, 1982).
Piaget, Jean. *Biology and Knowledge. An Essay on the Relations Between Organic Regulations and Cognitive Processes* (Chicago: Chicago University Press, 1972).
———. *The Principles of Genetic Epistemology* (New York: Basic Books, 1972).
Plata, S./Canaparo, C. 'Knowledge, Cinematography, and Space' in *The Cartographic Journal* (forthcoming).
Ricoeur, Paul. *Du texte à l'action* (Paris: Esprit/Seuil, 1986).
Romero, Francisco. *Ubicación del hombre: Introducción a la antropología filosófica* (Buenos Aires: Columba, 1974).
Rorty, Richard. *The Mirror of Nature* (London: Blackwell, 1980).
Sandbothe, Mike. 'The Temporalization of Time in Modern Philosophy', in *Time in Modern Intellectual Thought*, ed. by Patrick Baert (Amsterdam and New York: Elsevier, 1999).
Sarlo, Beatriz. *Una modernidad periférica. Buenos Aires, 1920 y 1930* (Buenos Aires: Ediciones Nueva Visión, 1988).
Sartre, Jean-Paul. *Being and Nothingness. An Essay on Phenomenological Ontology* (London: Meuthen, 1958).
Schrödinger, E. 'Discussion of Probability Relations Between Separated Systems,' in *Proceedings of the Cambridge Philosophical Society*, vol. 31 (1935): 555–563; and also vol. 32 (1936): 446–451. [1935]
Sloterdijk, Peter. *Critica della ragion cinica* (Milan: Garzanti, 1993).

Soja, Edward W. *Postmodern Geographies* (London: Verso, 1989).
Thom, René. *Paraboles et catastrophes* (Paris: Flammarion, 1983).
Vasconcelos, José. *La raza cósmica* (Mexico: Espasa-Calpe, 1976).
Virilio, Paul. *The Aesthetics of Desappearence* (Cambridge, Mass.: MIT Press, 1991). [1991a]
———. *Speed and Politics* (New York: Semiotext, 1991). [1991b]
———. *Art as Far as the Eye Can See* (London: Berg, 2007).
Whatmore, Sarah. *Hybrid Geographies. Natures, Cultures, Spaces* (London: Sage, 2002).
Wiener, Norbert. *Cybernetics or Control and Communication in the Animal and the machine* (New York: MIT Press, 1961).
Wright, Henrik von. *Explanation and Understanding* (London: Routledge and Kegan Paul, 1971).
Ziman, John. *Real Science. What it is and What it Means* (Cambridge: Cambridge University Press, 2000.)
Žižek, Slavoj. *Organs Without Bodies: Deleuze and Consequences* (New York: Routledge, 2003).

Chapter 2
Space, Geometry and the Colonies

According to the conditions of 'spatial thinking' laid out in the previous chapter, the idea of space has a direct relation to the *ancient* definition of geometry as the science of space or, more precisely, as the most comprehensive activity of thinking and imagining. Equally, as we will see later and as has already been discussed extensively elsewhere (see Canaparo, forthcoming), following the same perspective of 'spatial thinking', the notion of colony and colonialism have a direct relation to this *geometrical meaning* of the environment. That is to say, the understanding of the local condition of knowledge cannot be separated from the local construction of the space (environment) itself.

According to Herodotus (484 BC–*circa* 425 BC) the birthplace of geometry was ancient Egypt and its very beginning can be traced back as far as 3000 BC. Geometry, like philosophy, is an area that emerged in a *peripheral* space of the Greek empire. In its classical Greek formulation the word γεωμετρία can be translated as 'measure of the Earth', however Plato in *Philebus* was referring already to a 'philosophical geometry' that was different from a more practical activity of measurement.[1] More interestingly, in *Phaedrus* Socrates ascribes the creation of geometry to the god Theuth, who resided in the Egyptian city of Naucratis. This sense of an area/discipline/art that does not have a divine origin but cannot be found in the physical world either is a paradox that, surviving for centuries, and re-formulated by different authors, makes geometry and philosophy synonymous in their epistemic condition regarding the formulation of a concept of knowledge. This would become even more evident with the development of an open and non-axiomatic system in geometry. This began with the so-called 'analytic geometry' attributed

1 It is interesting to note that the origin of 'geometry' is not a discipline or area but an instrumental activity, as philosophy was understood as knowledge ('savoir') (see for example Foucault, 1984).

to René Descartes (1596–1650) who, not by chance, is also considered the first modern Western/European philosopher.

In a historiographical sense, *The Elements of Geometry* by Euclid (*circa* 300 BC) has been considered as the origin of geometry. Nowadays we know that Euclid wrote other more specific books about geometry in his time and also that he was not the first author to do so.[2] However, *The Elements* remains the only concrete work that has survived to our time.[3] In this context, if we consider geometry as the science of the *figuras* [figures] of physical space, then all spaces are 'geometrical spaces' because they require the construction of an environment and an *imaginary* point of view: all spaces are *figurative* by nature.[4] There is no space without figures – and 'figures' could be understood here as rhetorical figures, figures of thought or simple drawings. This was another aspect that would be transferred and made interchangeable within the philosophical activity: thinking (to think) would be, according to Descartes, for example, a way of creating concepts and establishing plans in relation to an individual subjectivity – a perspective that would be developed later, specifically in the twentieth century by Gilles Deleuze and Felix Guattari (1991). It is therefore paradoxical that Descartes himself, considered to be the modern Western philosophical model *par excellence*, was precisely the author who considered geometry (we must

2 Furthermore, the lexicological and historiographical analysis of Euclid's writings is very complex. And this is not only a result of the many translations of his writings but also because of the terminology deployed. Euclid, for example, used the word *aitema* ('request') which does not exactly match with the modern translation of 'postulate'. And this is relevant, as the reader can guess, in relation to the Greek notions of *doxa* and *episteme* and their relation with the modern idea of knowledge.

3 *The Elements* can be tracked down to at least the following origins: (i) *The Elements*, a Greek manuscript from the eleventh–twelfth centuries; (ii) *Compendio de geometría según Euclides*, an Arabic manuscript from the thirteenth century; (iii) *Opus Elementorum*, Venezia, 1482; (iv) *Eyclidou stoikheion*, Basilea, 1533; (v) *Los seis primeros libros de la geometría de Euclides*, Seville, 1576; (vi) *Opera Archimedis commentaria*, a Latin manuscript from the sixteenth century; (vii) *Opera*, Basilea, 1544.

4 The non-alphabetical writing of the pre-Columbian period has close links with this figurative condition.

remember that geometry was at the time concerned with the physical world) and philosophy as part of a global epistemological approach:

> He did not merely notice the overpowering certainty of mathematical proofs, and then attempt to use this to provide the basis for a new theory of knowledge. He did not merely use his mechanistic conception of the physical worlds as a way of putting forth his dualistic metaphysics statements that we have grown used to as characterisations of his major views are formulations based on what he thought he had achieved as a scientist and mathematician. (Popkin, 1983: 183)

From Pascal's (1632–1662) *De l'esprit géométrique* (*circa* 1657) to the developments of D. Hilbert (1862–1943) in his *Foundations of Geometry* (1902), the activity of thinking and the construction of a *corpus* in geometry have been divided into two tendencies: (i) one orientated to the axiomatization of concepts and (ii) the other orientated to free projections of a conceptual system. However, as a historiographical result, it is accepted today that any attempt to define the basic concepts of geometry is a waste of time due to its inevitable auto-referential condition and the *ad infinitum* nature of the arguments. This situation can also be tracked down in what we call 'philosophy' today. The artificial and fictional nature of geometry and the geometrical theories and systems are nowadays what make them close to philosophy, as used to be at the modern and classic origins of Western philosophical activity itself.[5]

Kant's definition of geometry as 'science de toutes les espèces possibles d'espaces' (*Gedanken von der wahren Schätzung der lebendigen Kräfte*, 1747: § 10) is indeed the immediate precedent and origin of this modern philosophical sense of geometry or, more precisely, of this entanglement of the philosophical dimension of the argumentation in geometry and the increasing *spatialization* of the concepts in philosophy. No less relevantly, the *totalization* assumed by Kant's concept of space points not only in the direction of the 'mathematization' of concepts about knowledge (A. Koyré: 'the mathematization of Nature'),

5 Already in 1820s Nikolai I. Lobachevsky (1792–1856), for example, called initially its geometric 'imaginary' (see for example the article 'Géométrie imaginaire', published in 1837).

but also in the direction of 'spatialization' of philosophical concepts or, more precisely, the 'spatialization' of any definition of epistemology (G. Deleuze/F. Guattari, 1991: 'géo-philosophy').

The development initiated by authors like Felix Klein (1849–1925) (see for example 'The Arithmetizing of Mathematics', 1887), Bernhard Riemann (1826–1866) (see for example 'Über die Hypothesen, welche der Geometrie zugrunde liegen', 1854), and Henri Poincaré (1854–1912) (see for example *On the Foundations of Geometry*, 1898) are a direct result of this entanglement. The most astonishing case is probably Poincaré who is still considered as a philosopher and also as a mathematician. The same could be said of Bertrand Russell (1872–1970) who wrote the stimulating and ever-useful *An Essay on The Foundations of Geometry*, in 1897.[6] Klein introduced a nomenclature in which geometry was organized into three parts: hyperbolic (Lobachevski), Elliptic (Riemann) and parabolic (Euclid). And this nineteenth-century taxonomy has its equivalent in the epistemology of Western/European philosophy's conceptual evolution as well: philosophy as conceptual matrix of disciplines (Nietzsche, parabolic), philosophy as main understanding of culture and culture evolution (Spengler, ellipsis) and philosophy as the comparative conceptual element for other conceptual systems (Cassirer, hyperbolic). 'The most important difference between all the geometries is that of the difference of philosophy.' (Osborn, 1950: 80). And, equally, the major epistemic difference between philosophies is that of the consideration of space: on analyzing Deleuze and Guattari's notion of nomadism in relation to Western/European colonialism, for example, W. Mignolo indicates precisely this lack of understanding of *another space* (see Mignolo, 2000: 72–88).[7]

6 This work of Bertrand Russell is also interesting because, according to the specialists (see for example B. Cassin, *Vocabulaire Européen des Philosophies*, 2004), he introduced the term *epistemology* to France.

7 And what's more, this connection and exchange between philosophy and geometry sometimes remains unnoticed by the specialists themselves. 'In *Philosophy of Geometry from Riemann to Poincaré* (1978) Roberto Torretti seems unclear as to whether he is writing a history of the philosophy of geometry or a conceptual history of non-Euclidean geometry' (Griffin, 1981: 375). And this is exactly what

However, the work that opened the door to the past in geometry – and made the entanglement with philosophy stronger – did indeed come from the Russian mathematician Nikolai Ivanovich Lobachevsky (1792–1856) who created the so-called 'non-Euclidean geometry',[8] grounded in a questioning of Euclid's fifth postulate, and proposing a new way to consider the 'geometrical space' (see for example 'A Concise Outline of the Foundations of Geometry', 1929). The more recent works of René Thom (1923–2002) in topology, the attempt to define the world geometrically, for example, would be unthinkable without Lobachevsky's perspective (see for example *Paraboles et catastrophes*, 1983).

It is as a result of these non-Euclidean developments that we can consider today two types of geometry: (i) the mathematical and (ii) the physical. And these match the two types of general philosophy that dominated our time: (i) philosophy in relation to the visible world (the period of natural philosophy, from the fifteenth to the nineteenth centuries) and (ii) philosophy in relation to the non-visible world (the period of scientific philosophy, from the nineteenth to the twentieth centuries).

Increasingly, with the evolution of the entanglement of geometry and philosophy, the notion of space becomes something varied and *local*. In the works of Descartes and Pierre de Fermat (1601–1665) on analytic geometry and also in those of Girard Desargues (1591–1661) concerning projective geometry, there can be found *en germen* the progressive idea that space can be studied without measurement or, more precisely, that measurement gives quantitative information about space but not qualitative conditions of it nor the characteristics and characters of it. This is exactly the same issue/problem that the concepts in philosophy tried to address as regards the immediate environment of the observer/thinker, from Baruch Spinoza (1632–1677) to Niels Bohr. The current derivation of this evolution in geometry – such as algebraic

should happen in a book about 'the philosophy of geometry', contrary to what Griffin seems to think.

8 Janos Bolyai (1802–1860) was also working at the same period of time in non-Euclidean geometry.

geometry, differential geometry, and indeed topology – are part of this tendency and historiographical evolution. And that explains in part why a contemporary European philosopher like Jacques Derrida, for example, was interested in geometry (see Derrida, 1962) and in the conceptual problems of philosophy (see Derrida 1973, 1980 and 1984).

The relevance of geometry, then, is not only due to the crucial role that it plays in defining the meaning of space or, more precisely, in defining the concept of space itself. Its relevance also comes from the fact that geometry and philosophy are two activities that work in a similar functional way. Their similarity and synonymous condition come not only from the fact that they deal with the same arguments, but also from a methodology (i) that works by constructing an artificial system, (ii) that is conceptual and (iii) that is considered as basic to understanding any perspective or conception of what we normally call world-view. The crucial role that cartographers, geometers and *agrimensores* play in the construction of the notion of 'territory' in Latin America, as we have already explored elsewhere (see Canaparo, 2005: 39–52 and 221–250; and also Canaparo, forthcoming: Chapter 6), is not casual and can be related to this Western/European intellectual evolution.

Considering the artificial nature of any conceptual system and assuming that any world-view is mainly grounded in hypothetical concepts (Popper, Lakatos) – because the concepts are hypothetical and conditional *per se* – is no longer an avant-garde perspective. It has become but the normal instrumental way in which knowledge is presented and distributed in public nowadays (e.g. market, academic community, etc.), especially when we assume, as we do here, that the Western/European notion of knowledge, hiding behind a 'universal' notion of science and truth, represents in reality a particular and well defined epistemology.

The common ground that geometry and philosophy share is the dimension of space or, more specifically, the creation and development of a spatial dimension. From a historiographical point of view, as we have already suggested, the Western/European perspective is dominated by this Greek-rooted/Greek-convened situation where the philosopher and the geometer confront the same issue from different points of view. While the philosopher concentrates on the *imaginary* relations of the physical world, the geometer focuses on the construction of physical

	Basic condition of knowledge	**Condition of evolution**
Space as Land	The local/locality as 'lugar'	The local/locality as 'sitio'
Space as Homeland	Imaginary	Distance Travel 'Lejanía'

TABLE 2.1. Evolution of space understood as land in local terms.

relations using abstract figures. If the philosopher constructs concepts to generate new meanings and understanding, the geometer is trying to construct the space in which these meanings and understandings will operate under a principle of reality. And this operative entanglement between philosophy and geometry is another reason why we postulate the notion of *geo*-epistemology as fundamental. It is fundamental not only for the *measurement* of things but also for the location, the placement, and the definition of concepts (communication, meaning, circulation, language) in relation to space (territory, place, environment).

Thus following this spatial condition, there are two aspects that we can now consider to be constitutive ground of the postulated 'spatial thinking' within the context of Latin America that we are exploring. One is what we will call the *epistemological dimension* and the other is a relation and evolution that we will describe as a *historiographical connection*. The first aspect is connected to the fundamental idea of *space as land* that became crucial to the definition of the 'local' itself. The second aspect is connected to the fundamental idea of *space as homeland* and gave birth to the basic meaning for the intellectual and imaginary activities within that locality, including terms like 'distance', *lejanía*, travel, and the imaginary. In this second chapter we will explore these two aspects according to what we consider their regular double formulation, as indicated in Table 2.1: (i) as a basic condition of knowledge ('condiciones de posibilidad', 'condiciones de hecho') and (ii) as a condition of the evolution of knowledge (meaning, concepts, *léxico*).

In Table 2.1 we indicated the combination between these two types of space with the two formulations each one has. And following from this we now conjecture that there are two possible formulations of what is local ('locality'): one understood as 'lugar' (place) and the other considered as 'sitio' (location). The first is connected with what some authors have indicated as *locus of enunciation* (see for example Kusch, 1970; Mignolo, 2000; and Mignolo, 2005); while the second is a function ('relación vinculante'). That is to say, the 'locality' is understood through the relationship between what is aboriginal ('aborígen') and what is the Western/European perspective or, more exactly, the sense of locality creates/encounters *the other* ('el otro': 'otredad interior' and 'lo otro': 'otredad exterior') in conceptual terms (see for example Dussel, 1975).

What the epistemological dimension and the historiographical connection also try to address is a particular relationship between land and writing that in our view is specific to the space that we call 'Latin America'. We have already discussed this with regard to the case of the River Plate area (see Canaparo, 2005). However, we believe it is a general phenomenon *that determines what Latin America is* and is not therefore a single 'colonial' aspect introduced by Western/European perspectives (see for example Adorno, 2000; Hill Boone, 1994; Mignolo, 1994b; and Leibsohn, 1994). The interesting and beautiful works of Gordon Brotherston (1930–) about the pre-colonial societies and communities make an accurate account of an already privileged situation given to the writing – in graphic, symbolical or alphabetical form (see for example Brotherston, 1992 and 1995). The mingling of this *native* condition and the Western/European perspective generates one of the possibilities of the above-indicated situation. The spatial condition that dominated writing as an argument or an architectural disposition (see for example Canaparo, 2005), is one of the characteristics that makes 'Latin America' different from other colonial and/or post-colonial spaces.[9]

9 In that sense the fact that authors and analysts present Latin America, not as a separate 'continent' but as an area that is an *extension* or *prolongation* of Europe, can also be explained by this situation.

1. The Epistemological Dimension

When considered from the point of view of 'spatial thinking' most of the usual terms deployed by authors and critics – such as 'language', 'culture', etc. – become *epistemological*. That is to say, they become part of a dimension where what is relevant is the way in which things become what they are, the way in which thinking becomes thinking, and we deal with the understand of understanding, the conditions of knowledge itself.[10] We will describe this as the *epistemological dimension* of our understanding of Latin America – a dimension that is basic to the comprehension of the evolution of ideas, institutions and concepts in general. More importantly, this epistemological dimension is the one that allows a more specific and less vague definition of 'colonialism'. In brief: colonialism in Latin America is not only the way in which knowledge is normally understood but the way in which it functions in relation to other concepts and, mainly, as a *hard core* or principal 'driving element' of a particular community or group. In this sense the works of Walter D. Mignolo have been very useful and refreshing, even when their proposed way forward is less interesting than their diagnosis (see for example Mignolo, 2000).

From the 'geographical' point of view (see for example Soja, 1989), just much as from the 'Latin America' point of view (see for example Castro Gómez, 1996), authors agree more and more on the fact that 'thinking spatially' is a way to understand and approach the 'second order of understanding', to use the cybernetic expression (see for example Foerster, 1995; and also Chapter 1) or the 'third

10 We have approached this question already in Chapter 1 under the notion of *cybernetic* – an aspect of contemporary Latin America considered as fundamental to understanding the way in which technology had changed not only the idea of science and the way that people consume and interact in society, but had also changed the idea of knowledge itself. The work of André Lebeau, *L'engrenage de la technique* (2005), as we mentioned earlier, constitutes a good general analysis of this situation; as does the argument deployed by Mignolo (2000: 115–124). And, even more specifically, the work of Santiago Castro Gómez, *Crítica de la razón latinoamericana* (1996), discusses the issue. From the perspective of 'geography' Soja (1989) also refers to this situation as a 'third space'.

space' to use the 'new geography' idea (see for example Soja, 1989). This epistemological dimension, then, is not a particular theoretical approach but is a way to try to see how the present understanding of Latin America – in terms of knowledge and cultural evolution – became what it is, conceptually speaking, and how this ensemble of notions and ideas managed to have the legitimacy that it has today.

All the crucial differences between Latin American – understood as a colonial space – and Western/European traditions are in this sense epistemological. They can only be described if we refer to the way in which concepts have evolved and, through the idea of culture and thinking have been transformed, implemented and developed by local institutions and authors. All of this also needs to be connected and explained in relation to a system of beliefs ('creencias', 'ideologías') and a system of references ('point de repère'). This is why a new *material* aspect has emerged recently in some of the analyses connected with the idea of Latin America. And this is also why we believe there is a connection between the ancient idea of *geometry*, as discussed – considered traditionally as 'philosophical' – (see for example Derrida, 1962 and 1990), and the 'post-industrial' and 'post-colonial' sense of *geography* (see for example Lévy, 1999). From this perspective we can hardly see any meaning in traditional analytical terms like 'reality' although we might be able to talk about *environmental reality*.

The epistemological dimension is a core, a set and/or ensemble, that deserves a proper and specific analysis. The epistemological dimension is where the concepts, objects and characters acquire their historical (as document, as legitimacy) and cognitive weight (as knowledge, as 'wisdom', as 'know-how'). It is not one place in particular but many and it is also the way in which some elements *function*. To dedicate time and energy to culture, comparative literature or local history without exploring this epistemological dimension is a waste of time or, more concretely, is a colonial exercise *alienado* (alienated) by the Western/European tradition of thought.

What we need to establish is an *architectural perspective* with regard to how the past – history, historiography, culture, etc. – has been established and determined: This is less a question of any particular subject or event than of the functionality they contribute and the way in which those subjects and events (through the perspective of history, culture, etc.) have been determined. The creation of the past, in this situation, is less a question about a particular matter than a question about how we understand the present, our relation to it and the role of the future (expectations, *horizon d'attente*) determined by this idea of a *constructed past*. We are not proposing a general 'revision' but rather an *epistemic switch* in the consideration of knowledge, a radical approach to communication, memory and language. The switch means determining and constructing an *environmental reality* for our notion of communication, memory and language.

Furthermore, the epistemological dimension attempts to acknowledge the impossibilities and limits considered from the spatial point of view and to find a way of approaching them radically. That is why we believe it is important to understand the relationship between 'philosophy' and 'geometry', even in the Western/European tradition, as a way of understanding and acknowledging the comprehension of *space as land*. The epistemological dimension is crucial for a constant and regular acknowledgement of the needs of a local conceptualization of the physical environment understood as *land* and, later on, as *world*. And that is also why there are three relations that, as we explored elsewhere (see Canaparo, forthcoming: Chapter 7), are crucial to the understanding of the emergence of Latin America as an 'autonomous' space or, somehow, as a 'new continent' (see for example Mignolo, 2005). These relations are (i) the definition of *lo familiar* in relation to the physical surroundings, (ii) the determination of the spatial limits in relation to physical and imaginary *fronteras*, and (iii) the establishment of a notion of the present ('el presente como determinacion de lo vivo, de lo que existe': Maturana) (see for example Maturana/Varela, 1980) and its becoming ('lo que sucede, lo que deviene y aún no tiene nombre': Perlongher) (see for example Perlongher, 2004) understood from the establishment of a notion of 'environmental reality' as a conceptual system that allows us to determine the previous two aspects, that is, a distinction between what is familiar and what is not, and an acknowledgment of the limits of our ordinary understanding and an extension of our semantic domain. If we believe that, for example, Mignolo's works constitute a very interesting diagnosis of the general epistemic condition in Latin America but a poor approach towards the future or the *posible* (possible), it is precisely because of this above condition: in order to be 'real', to be 'in touch' with what is Latin America in the *physical* sense, within a particular notion of present, the three conditions previously described are fundamental.

2. The Historiographical Connection

There is a direct historiographical connection between Western/European philosophy, the evolution of the idea and concept of geometry and the role that has been assigned historically to the colonies in relation to central imperial notions of knowledge.[11] From Plato (*circa* 427 BC – *circa* 348 BC) to Derrida, and from Aquinas (*circa* 1225–1274) to Husserl, there is a constant historiographical narrative that has been regularly re-enforced and re-visited at the moment when a new approach to Western/European philosophy has been is defined. Furthermore, from the Western/European perspective we can say that what we understand nowadays as 'philosophy' is a historiographical dimension and phenomenon. In fact, recent authors like R. Rorty and G. Deleuze have reacted against this condition as one of the main tasks in establishing/re-establishing a philosophical way of thinking (see for example Rorty, 1980 and Deleuze/Guattari, 1991). And it is indeed for the same reason that authors like W. Mignolo give relevance to the historical analysis that exhibits a distinction between 'philosophy' (universal, science) and 'thinking' ('indigenous thought', 'thinking style') (see Mignolo, 2000: 150 ff.).

And it is due to this basic historiographical condition that in the contemporary Western philosophical domain 'culture' is always understood as *historiography*. The book *Philosophy in History*, edited by Richard Rorty et al. in 1984, and the book *Philosophy and its History: Issues in Philosophical Historiography*, published by Jorge J.

11 In Chapter 1 we briefly discussed the connection between the idea and concept of philosophy and the notion of space, considered initially as geometry. Within the present Chapter 2 and in Chapters 3 and 4 we will expand some aspects of this connection.

E. Gracia in 1992, are almost perfect contemporary examples. Why is this situation relevant? Because in other non-Western/European latitudes 'philosophy', that is, a speculative domain established in a written format, does not rely on historiography but includes historiography as a necessary fiction and an element of present legitimization. And in these latter cases, when historiography is understood as the necessary fiction, 'culture' resists being formulated as historiography because it is entangled with many unsolved 'political' issues. In brief: at many latitudes, at the borders of Western/European domains and influences, *knowledge* cannot be formulated in a clear and unequivocal institutional way, as is the pretension within the academic and publishing markets in Europe and the USA. When historiography becomes and is considered as a ground in Western/European philosophy, within *borderlands*, on the contrary, it functions as a commercial and editorial instrument. This also confirms why 'naming' is a crucial epistemological aspect within Latin American space while in Western/European domains it does not appear so urgent.

The historiographical connection is crucial to understanding the role of *writing* within Latin America as a colonial space. Even more relevant is the comprehension of writing as a tool for establishing *space as homeland*. And that is also why it is relevant to study the entanglement between the Western/European perspective of writing-orientated knowledge and the local image-orientated idea of knowledge (see for example Pohl, 1994 and Leibsohn, 1994). In this sense it is relevant to point out that, within the pre-Columbian period, a non-Western/European *literacy* can be traced (see for example Brotherston, 1995; and also Adorno, 2000 and 2003), but this does not appear as 'culture' or 'thought' within the colonial Western/European approach to history and knowledge (see for example Cañizares-Esguerra, 2001). The *graphic dimension* that dominated pre-Columbian literacy was linked to the Jesuit idea of books and writing (see for example Feinghold, 2003), and the State sense of the administration of writing (the notion of document, etc.), all of which comes from the first period of the Iberian occupation. The notorious manuscripts, for example those of Juan de Santa Cruz Pachacuti (*circa* 1570 – *circa* 1650) (*Relación de antiguedades deste reyno del Piru, circa* 1613) and Huaman Poma de Ayala (*Nueva Corónica y Buen Gobierno*, 1616), are the result of this development/encounter between a graphic domain and an alphabetical order. In which case, as we will see when discussing the notion of *reversal thinking*, these works can be considered as the local historiographical and 'thinking' foundation, as is described, for example, in *América profunda* ('Deep América', 1963) by Rodolfo Kusch.

The space understood by these authors was indeed *visual* or, more precisely, the *space as land* was conceived as an *architectural design* which included many not entirely coherent elements. In this case, as already mentioned, the sense of 'local' was divided in two (see Table 2.1): the 'lugar' (place) and the 'sitio' (location). And this is what the aforementioned R. Kusch indicates in his *América profunda* as the two basic conditions of local knowledge and understanding: (i) the *estar aquí* ['to be here', 'to be at'] and (ii) the *ser alguien* ['to be someone'] (see Kusch, 1963: 25; and also Mignolo, 2000: 152). No less importantly, the *space as homeland*, was conceived as an imaginary construction (see for example Canaparo, forthcoming: Chapter 1) and as an environment circumscribed by three changing notions: distance, travel and *lejanía* ('far away', distant).

And it is here that our previous comment about the relationship between geometry and philosophy also becomes relevant. This graphic dimension and this entanglement carried out by authors like Juan Santa Cruz Pachacuti was profoundly grounded in the *non-representative* characteristics of the *mundo andino* ('Andean World') as well as in the linguistic structures of the quechua that clearly privileged a sense of the space, at least in the morphological and semantic sense. In brief: the conception of the space was 'geometrical' as it was founded on the idea that because the notion of representation was useless because the European naturalistic dimension is not locally relevant,[12] then the construction of space became an artificial local convention ('figurative' in the geometrical sense) of objects ('cosas') and people ('personas') and places ('lugares' o 'sitios'). And it is for this same reason that the

12 The most common approach for this perspective is the *figurativo* ('figurative') sense of drawing and painting. Yet more relevant is the basic belief that the physical world can truly be represented. With regard to this point see Canaparo, 2003 and 2007.

entanglement, between the Western/European convention of writing and the local graphic/design perspectives, becomes so strong that it is unthinkable to have one without the other (see for example Rama, 1993; and also Kusch, 1963: 17).

And, within this context, 'historiography' is then, or appears as, a set of concepts and values that operates almost as an *imaginary* (see Canaparo, forthcoming: Chapter 1). And if in Western/European approaches it works as a condition of the exercise of philosophy itself, in the borderland spaces it works as a limit of the present, as a point of reference that allows it to acquire some sort of local legitimacy. Under these conditions, the clear difference between the imperial space and the 'colonial space' lies simply in a different 'cultural' perspective. If in the Western/European tradition, the *continuity* of knowledge and understanding relies on a philosophical historicity, by contrast, within the colonial evolution the present, understood as a necessary continuity of things and concepts, becomes a formulation of space that is defined locally. And that is why there are three aspects that are always relevant in the evolution of this 'colonial space': (i) the relation between 'territory', the notion of property and the State (see Canaparo, forthcoming: Chapter 6 and Chapter 7); (ii) a specific dimension that we will call '*frontera*'s domain' (see Canaparo, 2005: 70–120; and Canaparo, forthcoming: Chapter 8); and, finally, (iii) another dimension that complements the previous two and which we will call 'environmental reality' (see Canaparo, forthcoming: Chapter 9).

3. Space and Colonial State

Less connected with a particular perspective about, for example, history or science studies in the Western/European tradition, the epistemological dimension is a way to re-consider the form under which the Latin American colonial world was conceived and, no less importantly, to explore its evolution and the conceptual consequences and developments of this original set-up. *Local Histories/Global Designs* (2000) by Walter Mignolo is probably one of the books, referring to Latin America as *subject*, that has recently insisted that the notion of epistemology be considered as a localized dimension of knowledge. Equally, as already mentioned, the historiographical connection is a consequence of this original set-up that has writing at its gravitational centre (see also Canaparo, 2005: 165 ff.).

Thus, as we have noted, it is 'spatial thinking' that prompts us to consider traditional colonial elements from another perspective. Or, more precisely, we are forced to consider certain elements under a constitutive status of the notion of space itself in the sense that these elements are no longer *characteristics* of a particular local condition but the description of the local condition itself. Also, again as mentioned before, two of these elements are the notion of land and the term 'territory'. Almost every society and country in the Latin American domain, mostly those which developed politically speaking within the nineteenth century, has founded its law system, its local history and its institutions on some form and/or formulation of the idea/term of 'territory', a situation that certainly has connections with what we can consider as a general Western/European perspective at the time (see Canaparo, 2005: 53–70 and 221–321; and also Canaparo, forthcoming: Chapter 6). And this Western/European perspective has already been explored by, among others authors, Michel Foucault who envisaged the conceptual consequences of the relationship between the land (home, city, countryside, territory), the body (population, anatomy) and institutions (State, school, etc.). At the end of the 1970s, in *Sécurité, territoire, population* (2004a), Foucault analysed the process of 'citizenship' as a *physical* dimension, the product of relations between imaginary concepts (*ratio pastoralis*: Foucault, 2004a: 233–239) and operative concepts (*ratio gubernatorial*: Foucault, 2004a: 233–239) or, more precisely, between concepts that function in an imaginary sense as well as in an operative sense – State, law, security, territory, population (see Foucault, 2004a: 3–30; and also see Foucault, 2004b: 3–28). A 'bio-pouvoir' was indicated by Foucault as the dominant condition of these emergent relations (Foucault, 2004a: 3–30). If it is pertinent to refer to Foucault's work here it is because it focuses on the *margins* of what was established conceptually during the very period in which most Latin American colonies were developing on a Western/European pattern. The initially *marginal* and less noticed conditions described by Foucault were to become most relevant at the end of the nineteenth century and during the entire twentieth century.

In *El imaginario Patagonia* (forthcoming) we have already examined the relation between territory, property and the notion of State as a fundamental principle for the development of local government and administrations from the nineteenth century onwards, at least for the specific sector of the so-called 'Cono Sur' (South Cone) (see Canaparo, forthcoming: Chapter 6 and Chapter 7). What we would like to briefly describe again here is the fact that the notion of property is also connected in its functionality with an intellectual and imaginary dimension. In *Tierras, leyes, historia* (2007) Rosa Congost discusses exactly the same point mainly in relation to the 'Península Ibérica'. The notion of property has continued to evolve and the idea of property as an unchanged concept of law ground in Nature is not a universal conception but a particular definition that corresponds to specific historical/historiographical interests or to a particular sociological context (Congost, 2007: 11–35). This aspect that might be considered irrelevant by some authors and critics became fundamental in the Latin American space where (i) economic production was traditionally related to various forms of exploitation of the land, and where (ii) the local *social distinctions* were mainly founded on this relation to the land. And, more recently, these aspects also became relevant due to (iii) massive emigration to the cities, (iv) the decreasing international and local value of agricultural products, and (v) the increasing concentration of land possession and economic production. However, as Congost (2007) himself points out, it was as a consequence of a particularly successful Western/European ideology of property that the idea of *universal rights* was introduced in many countries within Europe, and indeed within the colonies.

The fact that former colonies like Argentina, Brazil and Mexico became more and more dependant on importing of *divisas* and on the development of tourism is a clear result of this change in the recent consideration of property and land. The most important assets of the 'National' space are not localized in it – a situation that, as we have indicated elsewhere, is not new but is now emerging in its entirety and to having serious consequences (see for example Canaparo, forthcoming: Chapter 6 and Chapter 7). And if the countries of Latin America which were once considered to be more promising are in this state of *de-localization*, the conditions of societies like Equator or the

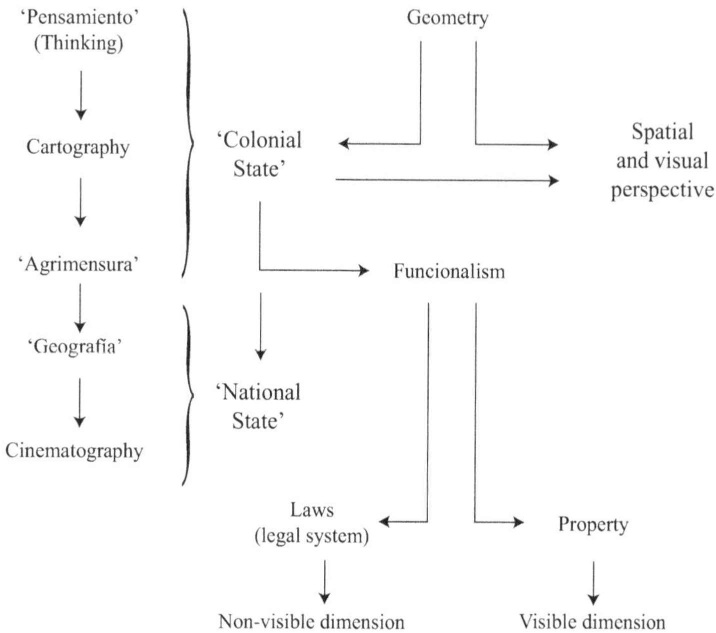

FIGURE 2.1 The evolution of space and the constitution of the dimension of a local State.

Dominican Republic, to name just two examples, where the above-mentioned elements have become the mainstay of the GDP, is much more extreme (see for example Thorp, 1998). The notion of property thus becomes confused and its traditional *material* sense of it is forced to negotiate its limits with the new more abstract, constantly changing and somehow technology-orientated notion of property. It is in this new 'non-material' sense of property that Foucault himself was interested in a result of a conceptual evolution within the Western/European perspective about (i) Nature/naturalism, (ii) the constitution of power/authorities, and (iii) the emergence of the self and the body (see for example Foucault, 2004a: 57–166).

In Figure 2.1 a summary scheme is presented of the situation already examined in *El imaginario Patagonia* (forthcoming) but also commented on here (see Canaparo, forthcoming: Chapter 6 and Chapter

7). The notion of 'thinking' is directly connected, in functional terms, with the local evolution of cartography (see Canaparo, 2005: 39–52 and 222–250; and also Canaparo, forthcoming: Chapter 2) and, in philological and cultural terms ('historiografía en sentido literal'), also appeals directly to the notion of geometry as discussed previously in Chapter 2.

The notion of 'thinking' evolves then through various stages of measurement and 'reality' and ends up as the constitution of a 'territory' as a way to determine a particular space. And it is later on, when this 'territory' starts to disappear, that the sense of 'geography' evolves towards a more abstract and 'imaginative' approach engendered, not by chance, by the emergence of cinematography (see Canaparo, forthcoming: Chapter 10). And, in fact, it is on the notion of 'territory' that the local constitution of the colonial State and the subsequent post-colonial State ('National State') relies. And, even when referring to a similar Western/European basis, that is, to both forms of State exercising the 'functionalism' indicated in Figure 2.1, there is nevertheless a terminological *switch* that it is important to address, that is, the modification of land considered as 'territory' and the absence of it ('post-territorial land') (see Canaparo, 2005: 53–70 and 221–321; and also Canaparo, forthcoming: Chapter 6).

Many elements, brought about by the development of the idea of 'territory', do not disappear but become secondary and subject to less visible elements (see Canaparo, 2005: 268–293; and also Canaparo, forthcoming: Chapter 6). Figure 2.1 tries to show how the concepts and terms that were indicated as relevant in the consideration of a particular perspective on the constitution of the local space, also become relevant in approaching the formation of the local State. More exactly, it tries to show how it is impossible to separate one aspect from the other.

And the colonial State, less relevant because of its doubtlessly growing administrative and political dimensions is also crucial because of its characteristic of/as *collector entity*. In brief: the colonial State functioned as primitive imaginary. And it is precisely when this functionality starts to disappear, as we have indicated already (Canaparo, 2005: 195–220), that the whole sense of land and territory comes into question. The colonial State as 'primitive imaginary' is the place in which people ascribe their sense of community, where objects

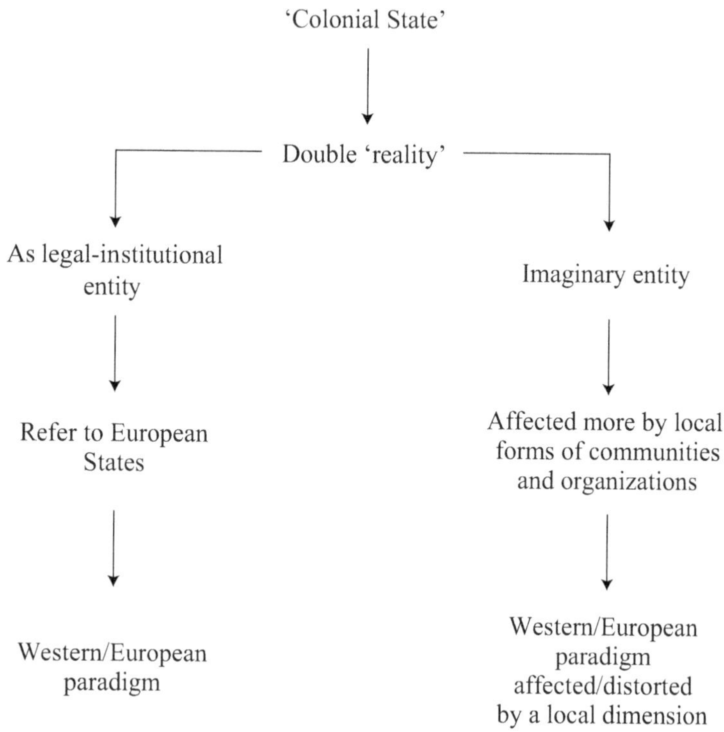

FIGURE 2.2 The double production of local sense/meaning.

get their approximate taxonomy, and where language gets confirmed by its dominant written status (see Canaparo, 2005: 221–250; and also Canaparo, forthcoming: Chapter 6).

Only later, at the beginning of the twentieth century, would this sense of State-as-territory start to be replaced by a much more powerful notion of cinematography (see Canaparo, forthcoming: Chapter 10). And after the disappearance of the local functionality of, for example, the notion of 'territory', or what we called *post-territorial land* (see Canaparo, 2005: 268–293), what emerges as fundamental is a reformulation of the *material dimension of land* (see Canaparo, 2005; 221–250). We will describe this reformulation as *place*, following the works of Edward S. Casey (see Canaparo, forthcoming: Chapter 6; also Casey, 1993: 3–40; and Casey, 1998: 331–342).

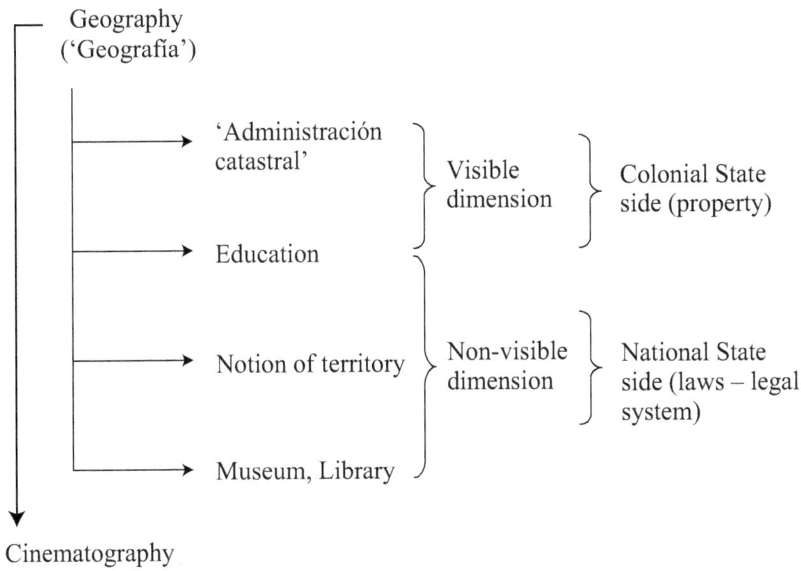

FIGURE 2.3 The 'geographical' dimension of the local space.

As already mentioned, the colonial State does not disappear but becomes transformed into a 'National State' (see Figure 2.1; and also Canaparo, 2005: 53–70), thereby transforming while at the same time sustaining the original condition of a 'double reality' as the dominant perspective, as described in Figure 2.2 (see also Canaparo, forthcoming: Chapter 7 and Chapter 8). Thus, a double production of sense/meaning will constitute one of the most salient characteristics of the local construction of 'culture' and/or 'world-view' (see Canaparo, 2005: 121–148 and 251–320; and also Canaparo, forthcoming: Chapter 8).

As mentioned earlier, the reason we are discussing the notion of 'thinking' as well as its evolution is because they are closely related to the constitution of a local space itself. Which is precisely why we brought into consideration the notion of 'geometry' and its constitutive aspects within the Western/European tradition of philosophy (or historiography of philosophy). It can be seen that, if observed closely from the perspective of the 'National State', the conceptual evolution of 'thinking' described in Figure 2.1, has two dominant components: the geographic (see Canaparo, forthcoming: Chapter 6) and the cinematographic (see Canaparo, forthcoming: Chapter 8).

In Figure 2.3 there is a graphic illustration of how 'geography' operates as a concept that mainly defines the entanglement between land and writing (or culture) (see Canaparo, 2005: 220–250; and also Canaparo, forthcoming: Chapter 2). Even though we have already considered this issue elsewhere (see Canaparo, forthcoming: Chapter 6), I still believe it is useful to point out once again that 'geography' describes an evolution from the administration of a topographic dimension (Cartography, 'Administración catastral') to a more comprehensive notion of land as totality ('territory') and land as a compendium of meanings ('Museum, library'). And that is exactly why, among other things, it is almost impossible to separate the local evolution of 'coloniality' (see Mignolo, 2000) and the development of 'modernity' (see for example Larrain, 2000). In contrast to what happens in Europe, the notion and implementation of the State in Latin America responds to a colonial need. And it is in this same sense that 'modernity' responds to the needs of the *créole* elites, who used it to explain the historical and cultural tradition of the 'newly-created Nations'. The creation of the State ('colonial' or 'national') at the local level cannot be separated from the colonial development and from the consequent emergence of concepts like 'modernity' that try to harmonize colonialism – and its local consequences – with a broader economic view conceived as universal organization and understanding.[13]

In this sense 'geography', understood as an extension of geometry (as described above), constitutes the native discipline *por excelencia*. And, as we have indicated already, the fact that the first local and general 'national' programmes of literacy and education were designed around geography is a clear indication of this situation (see Canaparo, forthcoming: Chapter 7). Since then, the exchange and interaction between a 'colonial side' and a 'local side' has become the natural

13 We understand this meaning of 'modernity' as the predominant definition of the concept when used in relation to a Western/European political system (see for example Orloff/Clemens, 2005 and Tuner, 1992).

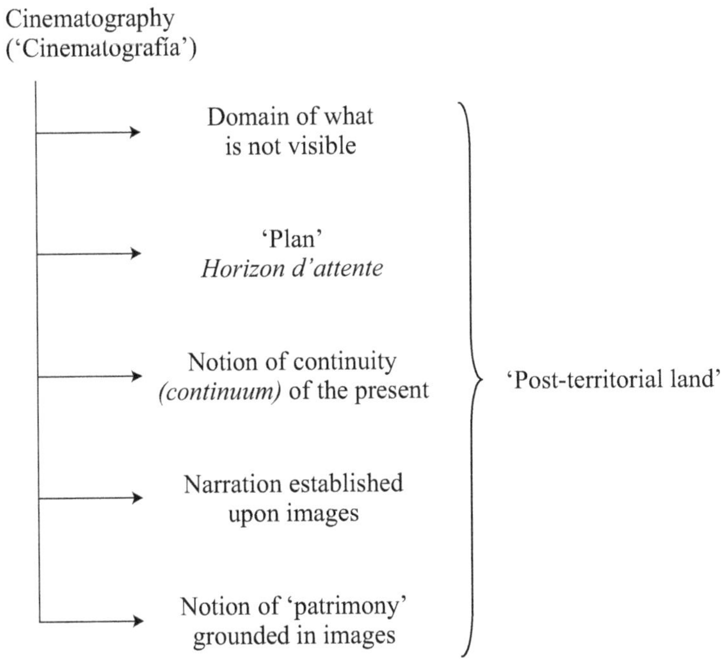

FIGURE 2.4 The 'cinematographical' dimension of the local space.

atmosphere of any local development.[14] And this why, when the notion of 'territory' disappears, local discussion of the *evaporación* (erosion, *disappearance*) of the land will become as important as concerns about 'globalization' and the introduction of a planetary-scale in everyday life (see Canaparo, 2005: 221–322). And this is also when the cinematographic dimension, which exists already, will become a fully operational perspective (see Canaparo, forthcoming: Chapter 8).

Figure 2.4 presents then a brief outline of the aforementioned 'cinematographic' condition and constitutes a simple enumeration of the

14 Figure 2.3 shows a diagram of the generation of a symbiosis between 'colonial State' and 'national State' during the nineteenth century. And also on how the local 'culture' was established in a process that evolves from the cartographic origins to the libraries and museums that conserve the 'patrimony'.

five main characteristics of what we call elsewhere *post-territorial land* (see Canaparo, 2005: 251–320). The introduction of a visual perspective and the abandonment of a naturalistic view, as indicated, make the notion of territory obsolete. The paradox of a culture whose idea of reality is founded on non-visible elements yet relies, for understanding it, on visual images, is one of the local aspects that foreigners find hard to understand. Ezequiel Martínez Estrada spent many years and pages trying to explore this random/quirky local condition of 'thinking'. And, given this condition of perception and understanding, the notion of the present as something that has continuity (*continuum*) becomes difficult to sustain because, increasingly, the present is understood as the expectation of what is possible and/or can happen in the future. The notion of 'plan' (in philosophical terms) and the notion of a *horizon d'attente* (in sociological terms) are at the core of what can actually be considered as 'local'. Furthermore, the very idea of *patrimony* connected to a physical entity disappears and, individually or collectively, a visual dimension becomes the natural environment for any sense of 'property' in terms of *values*.

Thus the 'cinematographic' dimension acknowledges the way in which a physical world becomes documented by what will be understood as 'reality'. Consequently, the collective meanings established through the *collector* condition of the 'national State' become obsolete from the semantic point of view. As demonstrated in *El imaginario Patagonia* (forthcoming), the 'cinematographic' dimension is less a specific imposition than a 'national', cultural and local evolution. This cultural evolution, moreover, moves from a principle of reality founded on the construction of a 'territory', and the whole system of signs and signals attached to it, to another environment where the land has been 'desertified' and all possible topographies seem to have become a question of memory (as conflict) or a question of the limits and misleading interpretations imposed by problems of *measurement* and/or scale.

Finally, by returning to the starting point of 'spatial thinking', we can propose (see Figure 2.5) a general perspective on the local, colonial notion of space – entangled with the creation of a State (see Figure 2.1 and Figure 2.2). And at the same time, *a particular* local functioning of 'thinking' can be addressed. The particularity of the status of Latin America or, more precisely, if we need to consider Latin America

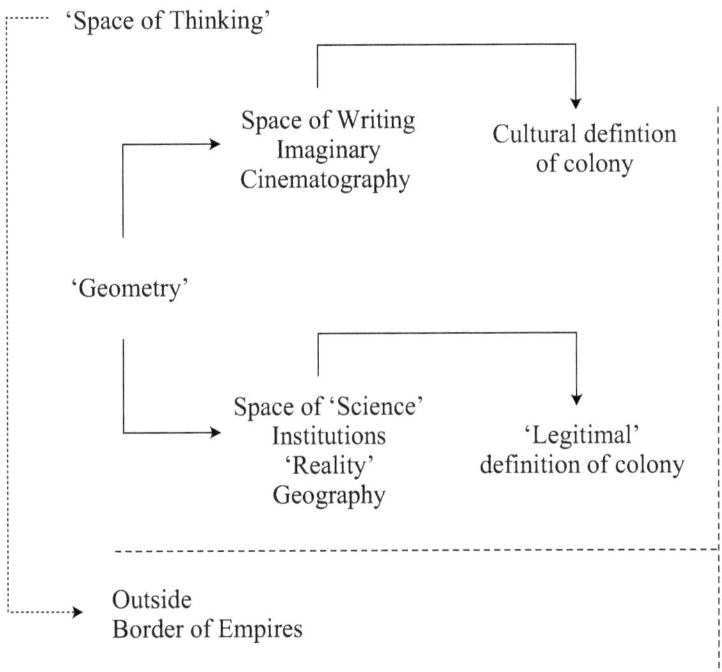

FIGURE 2.5 The 'space of thinking' (or 'thinking space') and its local development.

as a whole subject to be deployed, its truly epistemic condition, is then less connected to any folkloric or 'exotic' situation than to the condition of this 'spatial thinking' (see for example Pohl, 1994). It is not only the fact that Latin America is considered as a *geographical and cultural extension* of Europe, it is also the fact that the above-described 'peripheral' notion and relevance of 'thinking' (see Figure 2.1) become an aspect of distinction and reference.

The fact that many European authors consider the emergence of 'América' as a crucial historiographical *événement* in Western/European culture (see for example Todorov, 1987), at the same historical time that Western/European philosophers were trying to socially define their work *by concepts* (see for example Gracia, 1998: 36–51), is a situation that cannot be ignored and goes hand in hand with the comments made already about the role of 'geometry' in Western/European tradition. The need for the conceptual justification of philosophy as a discipline is directly connected with a geo-epistemic perspective about knowledge in such a way that it is almost impossible to separate the Kantian foundation of European philosophy and the spatial re-making that the 'discovery of the New World' represented for European culture. And, obviously, we are concerned less with the 'discovery' itself, than with the period of cultural European development from the sixteenth to the eighteenth centuries.

In fact, the 'cultural' emergence of historiography in the modern sense, that is, as a corpus of knowledge to which we can/must refer not only for questions of intellectual authority, first originated at the very time that the 'discovery of a New World' started to be incorporated by cartographers, botanists, and travellers into the Western/European domain. The fact that the incorporation of the 'New World' into Western/European knowledge and historiography happened just as 'philosophy' beginning to be justified as an area with its own history and without appealing to a metaphysical domain (see Gracia, 1998: 50 ff.), is not coincidental. The foundation of modern Western/European philosophy, the definitive establishment of the first European Universities as institutions that concentrated the available knowledge of a particular society, and the hierarchization of cartography as a technological and informative device, are elements that belong to the same geo-epistemological principle.

Figure 2.5 presents an outline of how the 'space of thinking', understood as 'geometry', works in relation to the 'double reality' mentioned above (Figure 2.2). On the one hand, 'geometry' is fundamental as a 'space of writing' – it functions as an imaginary or cinematographic perspective that generates a 'cultural' definition of the colonial environment. On the other hand, 'geometry' is generated as a 'space of science', in connection with institutions by a 'principle of what is real', that produces a 'legitimate' domain of the colony that is normally presented and justified as history ('historicidad'). In *El imaginario Patagonia* (forthcoming) and in *Muerte y transfiguración de la cultura rioplatense* (2005) we developed the relationship between what can be understood as 'culture' ('space of writing': Canaparo, 2005: 165–194) and what is considered as 'legal', authentic or legitimate ('space of "science"': Canaparo, forthcoming: Chapter 8). These two dimensions regularly interact and give rise to the most bizarre characteristics where sometimes the primarily colonial aspects get entangled with the more avant-garde and yet meaningless elements. The consideration, of the local space as both 'primitive' and 'post-modern' at the same time, a viewpoint mostly associated with a number foreign authors, is the most common result of this situation.

Rather than being an historical definition, this perspective of 'spatial thinking', and the conceptual evolution that it implies as a sort of *Begriffgeschichte*, generates an *oratio obliqua* of the colonial space with regard to Western/European philosophical tradition. This *oratio obliqua* can mainly be described as a perspective of space that is defined by the visual ('visualidad'). And it is indeed here that the aforementioned 'geographic' condition and its 'cinematographic' evolving form have their roots. We will explore this situation later on this work, under the name of 'reversal thinking'.

For decades the spatial analytical understanding of the colonial region was conceived according to the architectural structure of institutions that composed the State. In this sense all major local concepts – like 'history', 'ideology', 'society', etc. – followed this original spatial perspective. In many of the Latin American societies which developed during the nineteenth century, 'political science studies' – conceived as the maximum analytical expression focused on this architectural space of origin – were understood in two forms: (i) as the study of taxonomies or classifications of different forms of governability/governance (see for example Justo López, 1982), or (ii) as the study of the State or State-orientated institutions (see for example Lechner, 2001 and González Casanova, 1990).

The dominant trends of political Marxism, anti-imperialistic discourses and 'political engagement' perspectives that occupied the public sphere for almost all the twentieth century – *pace* military regimes or periods of long dictatorship – can be regarded as a consequence/complement of this 'double' original situation. The Western/European critical theory – from A. Gramsci (1891–1937) to M. Weber (1864–1920) – as much as the so-called 'local' perspectives – from the 'teología de la liberación' (see for example Berryman, 1989) to the 'anti-global' movements (see for example Galeano, 1998) – respond to the same colonial need: the 'theory' is a priority language that *instructs* societies and individuals (i) on how produce better descriptions and characterizations of forms of governability/governance, and (ii) on how the design and the understanding of a local State should help to improve/change it.

However, the strong introduction of neo-Marxist or neo-Leninist perspectives during the 1960s and 1970s (see for example Salazar Bondy, 1966 and Aricó, 1980) generated (iii) another type of analytical space perspective: the economic dimension. The evolution of concepts like 'history', 'ideology', 'society', etc., for example, within this colonial environment of economics and finance, show the radicalization of an already existent local situation: the spatial analytical understanding started to be translated from an institutional quasi-material design to another non-visible/non-material set of values and concepts. And it is this progress *towards the invisibility* that characterizes the evolution of analytical concepts within the colonial space of Latin America in the last few decades. However, the 'historization' ('historización') that characterized the traditional colonial approach (Western/European orientated) local 'political science studies', as mentioned above, has never disappeared and always emerges as a *colonial dimension* of the present.

4. Local, Far Away and Thinking

Many authors would agree that the first recorded philosopher of the fourteenth century in the Spanish domain at the time was not a Christian and his working language was not Castilian. Ibn Jaldún (1332–1406) (Abenjaldún) certainly occupies a marginal and peripheral condition within the history of the Spanish empire. The seven volumes of *Libro de la evidencia* ('Book of the Evidence'), from *circa* 1377, make Ibn Jaldún an author with a vision of the future not least because of his intellectual interest in the social aspects of the political communities of his time. The first volume of this monumental work is called *Muqaddima* ('Prolegómenos', prolegomena) and it contains the author's vision of a world understood as Universal History somewhat along the lines of the classic sense attributed to Herodotus (484 BC – 425 BC). More specifically, his political theory was that the history of the empires has a fixed pattern and is more directly related to internal aspects of the communities that support the empire and less to external factors. The axis of his analysis is the contrast between nomadic cultures and settled groups. Jaldún died one century before the encounter between the intellectuals and authors of Spain and Latin America. This missed chance of a dialogue between marginal authors working in a marginally defined domain of thinking is something that the diasporic movements and immigration exchanges of the beginning of the twenty-first century might now make possible (for a similar argument in contemporary European terms see for example Derrida, 1972 and Bourdieu, 1964).

And, once again, if we consider the role assigned within historiography to Jaldún, the definition of philosophy as a self-referential activity concerning the basic concepts of knowledge and understanding is somehow connected with a peripheral area, condition and person, with a *locus of enunciation* as we noted before. What seems to be considered as philosophy in Western/European terms always has a direct connection with the periphery of the imperial culture or dominant power that controls the institutions and the language that in theory contain such a concept (see for example Derrida, 1972). And, as noted before, it is this particular placement that in our view appears along with the emergence of America/Latin America, that is, the fact that the so-called 'discovery of the New World' helped to define the view in which Western/European authors were be perceiving/would perceive the notion of philosophy. Kant can only be understood in a historiographical sense, within the European context, if the spatial conceptual consequences of the 'discovery of the New World' are considered. Obviously, this did not happen because of any 'philosophical' influence of local authors but rather because America/Latin America offered a sense of open space, a sense of the far away, and a 'new world' that helped enormously to formulate a more accurate sense of modernity, universal philosophy and universal science. In other words, it helped with the first steps towards the establishment of a conceptual ground which could determine general values, a common language, and a system of political principles which could be implemented as a universal reality.[15] Those are the elements

15 Regarding this notion of 'reality' and its conceptual consequences see Canaparo, 2007.

that would be of prime concern for Western/European philosophers in the years to come. A similar argument, but deployed for different reasons and in different areas, has been explored in *Coded Encounters: Writing, Gender, and Ethnicity in Colonial Latin America* (Cevallos-Candau et al., 1994).

When space becomes the crucial element in the construction of a conceptual ground, the *locus of enunciation* (and not just the particular argument or content of a book or theory) emerges as the relevant issue (see for example Mignolo, 2000: 49–90). And that is also why a contemporary writer like Borges was effective where other so-called local 'philosophers' failed.[16] All the historiographies, systems of education, government policies and general ideas about culture, have been dominated locally by the assumption that what makes the local space and the non-local space (far away, paradigm, 'país', 'Nation') a single entity is the fact that it is considered as *an extension of Europe* and not just a simple colonial domain.[17] The geometrical-philosophical dimension already mentioned became relevant once again: the imposed Western/European dominance of the space became entangled with the local absolute relevance given to space in the perception of knowledge. The 'New World' became, at least in this strict sense, a *unique* case of planetary colonialism where the Western/European conquest was intended not only to implement some ideas but also to develop them further and reinforce the Western/European perspective itself. The fact that this *project* was considered 'cultural' and 'scientific' as well as political is what makes the case *unique*. And it is this *ambiguous* status in relation to Western/European institutions and culture that was at the origin of the subsequent future US financial and technological imperialism.

Since the nineteenth-century *créole* project of political independence from Western/European empires, communities in Latin America have been largely working around the creation of a *local tradition* of

16 With regard to how 'literature' can be considered as philosophy in the Western/European perspective, see for example Gracia et al., 2002; Gracia, 2001; and also Dubnick, 2002.

17 We have explored this idea and its derivations extensively in *El imaginario Patagonia* (see Canaparo, forthcoming: in particular Chapter 9 and Chapter 10).

thinking (see for example the classic Alberdi, 1852). The famous literary piece by J. L. Borges, 'Kafka y sus precursores' (Kafka and his predecessors) (see Borges, 1989: 710–712) exemplifies a paradoxical view of this. Formulated first in a more local way, for example in 'El escritor argentino y la tradición' (the Argentinean writer and tradition) of 1932 (see Borges, 1989: 267–274), Borges later managed to establish a *turn* on this formulation, a form of what we will call 'reversal thinking'. To indicate the cultural conflict and radical consequences that the *absence* of tradition in the Western/European sensibility can generate in a local perspective of 'thinking', Borges deploys the question *in Western/European terms*. He was not talking about Huaman Poma de Ayala or D. F. Sarmiento (1811–1888), he was referring to Kafka (1883–1924), a European Hebrew from Prague: the suggestion about the relevance of European culture becomes obliterated by the irony in presenting Kafka – a complete marginal of his own time – as a true representative author of the cultural tradition and dimension of Europe. Thinking from the space of the colonies, philosophising from the Western/European tradition, as we will see later, is the *motto* that can be used to express the first step of the 'reversal thinking' perspective.

And this is the first major *epistemic distinction* between the local environment and Western/European tradition. In local terms the 'tradition', the common ground in concepts and language *came from the future*, an agreement about expectations or, more radically, a necessary fiction *that needs to be written*. However, even in Borges' narration, there is a local element that is always there and becomes a sort of *common element*: the perception, the belief that, if any form of 'local thinking' is possible, it should be grounded on considering 'space' as a major speculative core. And it is there, within this common element, that a certain *regularity*, as previously mentioned, can be identified among the local authors and through historiographical narration. To be more specific, certain perspectives about space, or ways of describing and understanding space, became a common *point de repère* for those who would like to become 'authors' locally. The Inca Garcilaso, making up fake word genealogies and historiographies, or Huaman Poma de Ayala, constructing plans and charts of the Iberian Empire and its colonies, offer simple examples of what could be considered as an evolving local trajectory from the sixteenth century up to the twentieth century (on this point see for example Adorno, 2000; Leibsohn, 1994; Mignolo, 1994b; and in particular Brokaw, 2003). And this *trayectoria mutante* [mutant trajectory] is what we have already described as the *figurative* (space and form) or pictographic aspects of perceiving the environment, languages and culture (on this point see for example Brotherston, 1992; Marcus, 1992; Cevallos-Candau et al., 1994; Cummins, 1994; Mignolo, 1994a; Hill Boone, 1994; Zamora, 2005). This situation brings the local thinkers so close to the Western/European tradition, especially via the notion of 'geometry', but at the same time so far away, especially because of the impact of 'universalism', introduced, among other things, by literary taxonomies as well as an idea about science, in conceptual evolution, from Descartes and Kant up to the present time.

In the very synthetic and simplified summary presented in Table 2.2 we have tried to address the main aspects of 'spatial thinking' considered in the light of the arguments previously presented and in relation to the formation of an authorship in the local terms of 'thinking'. If we consider as local the authors that in a historiographical sense have not only been indicated as 'natives' but also as *defensores* ('partisans') of some local specific authenticity or identity in 'thinking', then the possible list cannot be reduced to the post-colonial period (after the political emancipation of the nineteenth century) but will be wide open to all periods since the arrival of the Spanish and before.[18] And this is exactly the point to which we would like to refer: within 'spatial thinking' the chronological divisions or time-scale scheme become irrelevant. In other words: what makes 'spatial thinking' particular is the fact that, even when chronologies and historical periods exists, in practice all aspects and periods *conviven* (are contemporaneous) in a *locus of enunciation* that authors normally refer to as 'present'. And this situation produces paradoxes in which, for example, elements assigned to authors in the sixteenth century also appear pertinent to

18 The consideration of the arrival of the Spanish as the *zero degree* of the analysis is inevitable due to the Western/European perspective of historiographies. However it is clear that a *conceptual construction* of the world that existed prior to the arrival of the Spanish is fundamental (see for example Cevallos-Candau et al., 1994; and also Brotherston, 1992). Of course, this construction will be, on the one hand, a historiographical narration and, on the other hand, a cultural object.

some contemporary authors. The idea of writing as *gnosis* in Huaman Poma, for example, is similar in historiographical terms, the only 'reality' available to us, to the approach taken by some nineteenth century writers, such as D. F. Sarmiento, and scientists like Eduardo Holmberg (1852–1937) even if the purposes and reasons of each one were radically different.

In Table 2.2, then, we have divided the local spatial trajectory, its continuity and regularity, into four constitutive periods. First there is the emergence of writing not only as an instrument or as part of a particular literacy process but as a proper *gnosis* itself (see for example Hill Boone, 1994 and Adorno, 2000). In this sense it is difficult to establish a distinction between 'thinking' and 'writing'. Secondly, the construction of a space that can function as world-view in semantic terms. Locally the Republic of Letters and the Republic in political terms become confused in many aspects. No less importantly in these circumstances, is the fact that writers and politicians constantly exchange their social roles, confirming, once again, the impact that authorship, as the most accurate form of writing in social terms, has locally. Thirdly, the establishment of a process of acquisition (or writing) and participation (in the political system) constitutes the first local approach to culture. And finally, the constitution of a local sense of knowledge subject to the elaboration of an independent and autonomous 'thinking', is considered as the latest achievement of the opposition to Western/European colonialism (see for example Mignolo, 2000). The idea of obtaining accumulative *derechos* [rights] that function to anticipate (*horizon d'attente*) the present runs through ('atraviesa') all the possible local notions of culture, or to be more precise, it becomes the definition of culture *per se*. It is interesting to notice how local authors like Huaman Poma and Western/European authors like Fray Bartolomé de las Casas (1484–1566) encountered the same issue of 'rights of the aborigines' but in one author this refers to the activity of writing (and/or producing physical/cultural objects) and in the other to a theological perspective about Nature. We will return to this point later in Chapter 4.

And it is here that it becomes pertinent to mention that the four periods indicated in Table 2.2 are *not historically grounded*: these periods constantly exchange elements and forms, and prevail in the present under different dimensions. They did not disappear to absorbed

Author/ Authorship	Dimension	Horizon d'attente
Huaman Poma, Inca Garcilaso, Concolorvo, Sor Juana	The constitution of Writing	The 'right' to writing
D. F. Sarmiento, J. Martí, José Carlos Mariátegui, Juan José Alberdi	The constitution of the Republic space	The 'right' to a political system
E. Rodó, Vasconcelos, E. Martínez Estrada.	The constitution of cultural space	The 'right' to culture
J. L. Borges, L. Zea, Salázar Bondy, E. Dussel, Eduardo Holmberg	The constitution of knowledge	The 'right' to 'pensamiento' ('thinking')

TABLE 2.2 The local non-temporal trajectory from the 'spatial thinking' point of view and in relation to a dimension of authorship.

by some new formulation but constantly change and adapt themselves. By the same token it is also pertinent to point out that the notion of *derecho* ['right'] presents an important semantic diversity that is also attached to the scheme in Table 2.2 and to the very idea of a trajectory in 'spatial thinking'. *Derecho* can be understood in six different ways: (i) as the right to an object (property) or thing (principle), (ii) as something to which we are entitled ('derecho escrito', law), (iii) as the 'right' direction, as the most efficient and correct way to address a problem or to address our body ('avance', 'adelante'), (iv) as a corpus of decisions ('disposiciones') that can be understood in two ways: like historiography or like 'jurisprudencia' (jurisprudence, narration of leading cases, digest), (v) as a definition of a particular type of law or implementation of the law, and (vi) as 'derecho inalienable de las personas' (natural right, human right). The four formulations of 'rights' that appear in Table 2.2 then need to be contrasted, according to the period and particular situations, with these six different meanings which originate from Western/European tradition. The result is a multiplicity of situations that, considered as a whole, provide a framework to locally define what we understand today as 'culture' in Western/European anthropological terms (see for example Sahlins, 1978 and 1995; and also Geertz, 1977 and 1985). In this sense 'Latin America' is a physical dimension ('continent') that is yet to be discovered.[19]

'Spatial thinking' then functions as an *environment* ('ambiente-escenario-amueblamiento'). And this environment is understood here in at least three ways: (i) as 'ambiente', that is, as a physical space but also as equivalent to a particular measure of space and to a determined and closed space; (ii) as 'escenario', that is, as a determined 'set design' (*sceneggiature*) (objects, things, setup, scenario), but also synonymous with *stage* (bodies, people, public space, exhibition); and (iii) as 'amueblamiento', that is, as the distribution of physical objects in a particular established space and/or as imaginary-possible ('mondi possibili') the distribution of objects and people in a determined/defined space (see Table 2.3).

Within the context of 'spatial thinking' the only effective 'reality' – in the sense of *realitas* (see Canaparo, 2007) – that can be referred to is an 'environmental' one. The notion of *environment*, as a constructed world-view and as a relationship to the immediate physical system, is *holistic* enough to comprehend the multiple condition of what is understood locally as 'present' and which seems to be a regular concern among those authors who define locally the sense of 'thinking' (see for example Zemelman, 1989 and Medin, 1983).[20] And if any possible principle of reality needs to be *environmental*, equally, the only available sense of 'global condition' – as synonymous with *environment* – is a *constructed* one. The conjecture that concepts, the semantic construction of our physical surroundings, and the elaboration of a sense of *world* are aspects which are not only interconnected but which also *become a developed ensemble/assemble* is a very well-

19 The increasing amount of research conducted in the area of so-called *material culture* responds, I believe, to this situation.

20 The relation between 'pure becoming' and language was one of the main problems considered by European philosophers during the twentieth century (see for example Derrida et al., 2001). However local authors have been concerned with a more historical-orientated issue where what remain to establish is the notion of *evénément* as a desirable combination of (i) accurate information, (ii) legitimate political status, (iii) minimum epistemic conditions (see for example Lechner, 1981 and González Casanova, 1981).

known Piagetian perspective (see for example Piaget, 1937 and 1967; and also Glasersfeld, 1995). The basic relevance of this perspective here is the fact that this *constructivism* is the only way by which a sense of environment can be constructed and used locally. This is a perspective that is also connected with non-local authors who somehow postulate a sort of 're-invention of Nature' as a form of epistemic challenge to the contemporary naturalistic conceptualization, as Donna Haraway does in her classic book *Simians, Cyborgs, and Women* (1991).

The fact that there is a direct relation between (i) a certain notion of 'private property', (ii) a particular functional view about technology, (iii) a *financial* perspective on values and goods, (iv) an abstract and non-visible sense of matter ('materialidad'), and (v) the evolution and implementation of the State as the crucial institution in public affairs, is something that in the past was curiously not connected with an idea of *environment* (see for example the classic Nicholson, 1973). Of course, it does not seem surprising that colonial/local authors were somehow already addressing these relations as crucial and necessary elements in defining a local sense of environment (see for example Torres García, 1938 and 1948; Sábato, 1951; García Canclini, 1999 and 2001; and Murena, 1961). In this sense it is always interesting to see how something that is considered as an intellectual or cultural result of Western/European development and evolution – something that we can identify as post-modernity, post-rational philosophy, deconstructivism, post-industrial society, etc. – becomes associated with a 'post-colonial' perspective by local authors. That is to say, it becomes associated with a new attempt to reformulate the way in which the local space is defined (on this subject, see for example, Mignolo, 2000: Chapter 1).

It is only recently, in works like *Changing the Atmosphere: Expert Knowledge and Environmental Governance* (2001) or *L'engrenage de la technique* (2005), for example, that these relations have been considered as partially or totally relevant in the foundation of an integral notion of environment by Western/European authors (see Lebeau, 2005; and Miller/Edwards, 2001). In exploring the idea of *propiedad* (asset, property, possession), Rosa Congost (2007), to give another example, showed how it has evolved and changed over different periods and how the idea of *propiedad* as a single and universal concept is a perspective that emerged recently in a Western/European context, especially with the institution/institutionalization of the local State in the Weberian sense. In a similar way the previously mentioned André Lebeau, exploring the consequences of contemporary and global technological development, exposes the perspective of a single and universal notion of *environment* (as Nature or philosophical naturalism) as a specific perspective designed to carry out a particular Western/European notion of scientific purpose (see for example Schäfer, 1983 and 1993). Even recent traditional works, like *Patterned Ground* (2004) have begun to question a unified and single conception of what is still recognized as 'nature': the entanglement between culture and nature is such that conventions and language issues prevail over any scientific theories (see Harrison et al., 2004: 15–25 and 26–32).

In this sense it is somehow inevitable that the notion of environment has become unavoidable in relation to the idea of *actualidad* (contemporaneity, *aktualität*) because it contains and addresses a number of elements that are constitutive of the very notion of 'present' as pure becoming. Earlier, when we referred to *environmental reality* it was to the situation described above that we were alluding: the notion of *environment* is actually the one that allows us to compile not only a semantic diversity but also to contend with the fact that many concepts are mainly 'functional' and that their status comes less from some static description/definition than from the varied nature of their evolution.

In Table 2.3, we list the three forms by which we exposed environment as 'spatial thinking' and associated a number of characteristics with each form. In doing so, we are trying to determine a possible generic formulation of the issues (concepts, situations) that an environmental reality, as discussed above, should address. Understood as 'ambiente' the environment as spatial thinking can be considered (i) as what is immediate or far away, (ii) as a sort of globalism founded on a set of close functional elements, or (iii) as a principle that allow the coordination of a variety of elements and situations. Understood as 'escenario', the environment can be considered, in terms of spatial thinking, as (i) the need to regularly establish conditions as a set of situations, (ii) the place where understanding and explaining are synonymous, and (iii) a place where single elements do not have 'life' of their own. Understood as 'amueblamiento', the environment can be considered, in terms of spatial thinking, as the place where technology

Element	Characteristic Aspects
'Ambiente'	• Immediacy and the 'far away'. • The sense of 'globalism' is founded on the functionality of particular elements. • The dominant sense of *atmosphere*, in its original Greek sense of spatial definition, as the conjunction of 'air' and 'sphere'.
'Escenario'	• Principle of Coordination. • Conditions are not explained singly, it is a set of situations that need to be taken in account. • The understanding of the environment is no different from the explanation we give about it. • The dominant sense of *mise-en-scène* or *mise-en-narration.*
'Amueblamiento'	• One element does not correspond to another single element but to the whole situation and/or condition. • The perception of immediacy, the use of technology, and the value-determination of objects are part of the same process of knowledge and understanding. • The dominant sense of *composición* [composition] as a physical and imaginary arrangement.

TABLE 2.3. The possibilities of the spatial thinking stand point understood as environment.

and the perception of ordinary life are not separated. If the 'ambiente' gives us a spatial thinking perspective in the form of *dominio atmosférico* (atmosphere), then, the 'escenario' works as a *mise-en-scène* of the spatial thinking and the 'amueblamiento' as a *composition* of the spatial thinking.

If our hypothesis is correct, this notion of environment offers the only locally available possibility of generating a certain efficacy with a principle of reality or, with concepts that will function as a principle of reality (*realitas*), under a colonial or postcolonial condition (see Chapter 3). Therefore, when we refer to an environmental reality what we are trying to address is the specific combination of 'ambiente', 'escenario' and 'amueblamiento' that characterize a period and/or situation from a local perspective without losing the determination and limits imposed by general conditions.

In these particular circumstances, the notion of environment is also expressed through other concepts that can deliver a similar sense of totality and composition in the way that environment does. Two examples closely related with our subject are the ideas of *weather* and *meteorology*, both of which are connected with the notion of *forecast*. That is why authors concerned with criticism or critical studies tend to use concepts that work in this 'weather dimension' when referring to the 'global' trends (see for example Ross, 1991). Furthermore, those analysts concerned specifically with the environment in the sense discussed here understand the epistemic dimension of it: 'Meteorology as Infrastructural Globalism', an article by Paul N. Edwards published in 2006, is one of the best recent examples of this perspective (see Edwards, 2006). On the other hand, when very well known commercial authors like Frederic Jameson or Naomi Klein use the word 'globalization' or 'ecology' they are also trying to address this need to deploy spatial concepts able to 'absorb' all the elements present in such space (see for example Jameson, 1998). Yet more relevant is the fact that the increasing *density* of the concepts that name/refer to space goes together with the recent development – and trendy tendency – of general perspectives such as 'globalism', 'transculturalization', 'planetarization', etc. And when, for example, specialists in science studies try to generate a sociological model for the development of science and to explain its conceptual evolution, or when social scientists address the loss of boundaries between disciplines, authors once again attempts to address a similar situation. This situation can be defined in terms of (i) a non-soluble conflict between the local and the global,[21] (ii) a high density of elements per concept, and (iii) a condition of prediction (*horizon d'attente*) as fundamental to the characterization of the conceptual present. Finally, radical political views from authors like Robert Nozick

21 The traditional problem is called a 'hermeneutical circle', which occurs when an explanation grounded in totalities or in its parts just vanishes when we realize that the conceptual situation presents totalities and particularities not as alternatives, but as existent parts *in conflict*.

in *Anarchy, State and Utopia* (1978) or Edward S. Herman and Noam Chomsky in their book (1988) and documentary (1992) *Manufacturing Consent*, also addressing this issue of environment in an *atmospheric way*. We will return to this condition/situation in Chapter 4.

The notion of reality ('el principio de realidad') attached to the concept of environment is relevant not so much because of any specific or particular aspect of the meaning of environment that we are exploring here. It is relevant, rather, because of the functionality and capabilities that the *environment* itself has in determining/naming space (territory, land, imaginary, etc.) as a linked and dense composition (*compositio*, taxonomy) of elements. It is not a matter of chance that we discover the same semantic possibility in the notion of geometry described at the beginning of the chapter. And it is not far away from the pictographic sense of space that specialists attribute to the pre-Columbian communities (see for example Brotherston, 1992).

That which constitutes the basis of reality is the 'ambiente' or 'environment'. And it is precisely in connection with this condition of a 'principle of reality' (*realitas*) as environment that the question of immigration/emigration and diaspora acquire unexpected relevance. Immigration and/or emigration, as concepts, do not refer only to the movement of people but also – and no less significantly – to a location or dislocation of 'cultures' and languages (see for example Ortiz, 1995, 1998 and 2005; and also for a more general approach Bhabha, 1994; Said, 1994 and 2000; and Rushdie, 1991). One can point out that these are Western/European concepts of immigration/emigration, and that is correct. The starting point of these concepts, their 'origins', is an inevitable Western/European experience from the point of view anyone in Latin America. What makes the difference, as suggested before, is the *transformation* these concepts undergo in the colonial environment and which enables us to describe them as post-industrial societies/communities. Abdelmalek Sayad's (1933–1998) classic work on immigration is a solid confirmation of this contemporary 'cultural' definition of immigration/emigration (see Sayad, 2006a, 2006b; and also Bourdieu/Sayad, 1964). If in the past there was a market for people closely attached to the reasons and explanations of immigration/emigration, equally, and more extremely evolved, there is nowadays a general sense of market that covers almost every human dimension, from

Environmental Element	Dimension in terms of emigration/immigration
'Ambiente'	• Immigration and emigration movements remain constitutive in cultural terms of the communities indicated as *displaced societies*. • Because of the relevance that the notion of immigration and emigration has in the evolution of the local culture, both were notions regularly interchanged and later became confused as their original meaning started to be constantly reformulated in such a way that in practice there was almost no distinction between them. • Multilinguism and traditional *diglossia* are not always the result of a culture founded on immigration/emigration. It can also happen that the multilingualism or *diglossia* situations are transferred to a cultural level, not only as a 'language' phenomenon, creating a sort of completely schizophrenic culture or cultural perspective. • Lost of roots in the physical sense and therefore lost of legitimacy of any 'monumentalism' in relation to identity. • Condition of *movilización total* (total mobilization) to use the Western/European war terms of E. Jünger.

TABLE 2.4A. The immigration/emigration condition in the light of the local environment considered as 'ambiente'.

clothes to ideas, which is related to the ideas of immigration/emigration. Immigration/emigration as concepts are an constitutive/integral part of what we call 'Latin America', not only for the obvious historical reasons of constitution of the local population from the fifteenth century up to today, but also for a series of non-temporal elements, some of which are indicated in Tables 2.4a to 2.4c.

Immigration (*immigrare*) and emigration (*emigrationis*) are 'spatial' concepts *par excellence*, in fact the semantic distinction between the two is normally based on a consideration about specific space.[22] While immigration is defined in relation to a reception space (the space in which we are arriving), emigration is related to a space where we no longer are (the space that we are leaving, the space that is left behind). We immigrate *towards* a particular space, we emigrate *from* a particular space. Nowadays, all the aspects of what we understand as culture in the anthropological sense in Latin America are affected by these two concepts, a situation that has not always been considered as relevant by the authors and critics of colonialism and the post-colonial situation (see for example García Canclini, 2001 and Mignolo, 2000). In fact, what can be considered as the radical change inflicted by the Spanish to totally indigenous populations, paradoxically, are not only to do with a military defection but also with the condition imposed on the local environment by the colonization driven by *alien bodies* – individuals, objects, artefacts.[23]

Therefore we should not only use the terms 'emigration' and 'immigration' to refer to individuals but also to bodies. Furthermore, we should not think in terms of static exchange but in terms of *travelling* (on the subject see for example Pratt, 1992). And that is why, when referring to the local notion of environment, we emphasize its *artificial* aspect as opposed to a traditional anthropomorphic or scientific-orientated notion of environment. Nature, as philosophers like Jean Piaget and scientific analysts like Donna Haraway (1944–) have shown, is a construction, an artefact, and not an analogy of the physical world (see for example Piaget, 1967 and Haraway, 1991). The actor-theory from Bruno Latour, for example, is one of the recent developments which, from a Western/European point of view, attempts to approach environment from the *atmospheric* perspective already mentioned (see for example Latour, 2005; and Latour/Weibel, 2005).

22 Obviously there are many forms by which this 'space' can be understood. We will return to this subject later, in Chapter 4.

23 Immigration/emigration can be subsumed/merged into the idea of *migration* as far as we understand this in the extended sense already discussed, that is, where we refer not only to the movement of people but also to that of objects, technologies, animals and natural resources. Rather than living in a transitory/transitional or global environment we live more and more in a *migratory planet*.

The settlement of objects, buildings, roads, ports and human beings, produced by Europeans in 'Latin America' between the fourteenth and nineteenth centuries, can indeed be considered in the broad sense of movements of emigration/immigration, but less as an historical fact and more as a non-reversible evolution of the environment. The present nomadism and de-localization of thousands of inhabitants or entire communities in Latin America, contrary to appearances, is not a new phenomena but something that goes back to a spatial perspective and knowledge condition of pre-Columbian communities and indeed to the process of colonization itself generated by Western/European powers (see for example Mignolo, 2000). What was the environmental reality for the aborigines, as a consequence of the brutal conquest carried out by Western/European powers is, paradoxically, now a reality for the so-called natives or *criollos*, the descendants of those first conquerors (see for example Mignolo, 2000). That is what populist local authors like Eduardo Galeano consider, from a fatalistic perspective, as a sort of inevitable 'destiny of Latin America' (see for example Galeano, 1973).

Several interesting works have been dedicated to the subject of immigration in Latin America (see for example Schneider, 2000), but never to emigration, especially not the so-called 'internal emigration'. However, our purpose here is not to survey this area or cover these deficiencies. Our aim is merely to situate in general terms the local notions of immigration/emigration within the context described so far and as functional concepts within the context of the location of knowledge. However, two or three general considerations are necessary. Given the characteristics of immigration/emigration specified earlier, we can assume that all immigration/emigration is *diasporic*. Because in one form or another the point *from* which nowadays we immigrate/emigrate always *disappears* in the traditional sense of 'roots' or in the archaeological/ethnographical sense of 'monuments'. The 'planetarization' of culture, the extreme colonial perspective of 'one culture', in which scientific reasons, market reasons or geographical reasons are intermingled, produced the paradoxical effect of turning every individual into a sort of member of a global diaspora. This situation is, of course, more evident in peripheral areas of the Western/European domain. And this is why what were previously understood as 'diasporic populations' are today under conceptual threat.

Environmental Element	Dimension in terms of emigration/immigration
'Escenario'	• Communities and populations are 'constructed' around movement of immigration and emigration. We will generally label them *displaced societies*. • However within these *displaced societies* there are degrees of displacement: 'aborigines' and 'natives' do not relate in the same way to the space, just as 'internal' immigrants/emigrants differ from 'external' immigrants/emigrants. • The movement of immigration and/or emigration also get mixed up with the general movement of *values*: the incredible amount of currency for example that is sent to Latin America from Europe and the United States is not a new phenomenon, even if its characteristics are *sui generis*. What is new in this sense is the dependency of the local State on these values and their movements. • The extreme mobilization of assets: objects and values become dominated by *financial* characteristics. The elements of traditional social identification become *commodities*. Consequently, the notion of *agency* starts to be considered as a fundamental social mechanism within the communities.

TABLE 2.4B. The immigration/emigration condition, considered as 'scenario', viewed in the light of the local environment.

Another issue to consider in this context is the disappearance of the notion of *distance* (see for example Canaparo, 2005: 107–120 and 205–220), because it completely changes the sense of immigration/emigration. For instance, the aspect of *non-reversibility* that characterized the notion of immigration/emigration at its classical origins is no longer present. And in Latin America nowadays people do not immigrate/emigrate *completely*: somehow in a physical or imaginary way people keep a 'territory' in their original place – even when this *no longer exists*, physically speaking. And that is also, as we have already shown, why the sense of *place* changed radically in relation to how and where the spatial and geographical conditions, under which we read the political dimension of Latin America, were realized/developed during the nineteenth and twentieth centuries.

And, furthermore, the characteristics of Latin American *as an extension of Europe* were also modified by the disappearance of the notion of distance and by the changing perception of what 'place' is. It is only now, in the twenty-first century, that what we call Latin America stops being considered as *a geographical extension of Europe*. And it not only happened because radical changes occurred to the same notion of *geography* (see for example Edwards, 2006 and Canaparo, forthcoming) but also due to the fracturing of an attached Western/European notion of space that is no longer functional in local terms. The increasing number of studies about pre-Columbian periods, about colonialism and 'globalization', are certainly a consequence of this situation. Paradoxically, and without any military or revolutionary State-orientated action, it was under this above described situation that, from the cultural and 'epistemic' point of view, the *independence* from Europe started to happen. The fact that the implicit understanding that 'follows' the Western/European model of social and economic development is not an available possibility anymore goes hand in hand with the belief that such a model cannot be produced locally. This is something that, curiously, even radical colonial critics such as Walter Mignolo, to mention but one example, did not realize (see Mignolo, 2005). We will return to this point in Chapter 4.

Within the context of immigration/emigration, the notion of *population* also acquires a different perspective: seen from a long term perspective large agglomerations in Latin America remain exactly as they were 100 years ago and, furthermore, they have been dramatically increasing in size. However, looking more closely, these large agglomerations are always *in movement* 'internally': only exceptionally do individuals remain in the same urban space. The idea is that a *habitat* needs to be re-constructed several times in one individual's lifetime, and, in many cases, is even more in constant construction/re-construction (see for example Sennett, 1998). The relation between quantity and movement, in terms of population, has changed radically since the middle of the twentieth century.

Finally, we need to contrast the above-mentioned characteristics of environment in relation to the general 'cultural' aspects of immigration/

Environmental Element	Dimension in terms of emigration/immigration
'Amueblamiento'	• These *displaced societies* no longer have a common topographical/land site. Their 'territory' is cartographic but not material in the general sense of 'being familiar'. And the material references are always locally determined, mainly in cities/urban domains. • The speed that dominates the institutional functionality of our societies and the omniscient relevance of the flow of information (Virilio: 'dromology') creates a sense of movement that is constantly matched by the displacement caused by the emigration/immigration of objects, people and ideas. • Despite the hypothesis of the seminal works about immigration/emigration (see for example Sayad, 2006a), the issue of 'identity' is less and less relevant. When mixed with a perspective on colonialism, the emigration/immigration situation of individuals results in an issue of subjectivity, i.e., in a process of making individuality the *subject* of a particular local dimension. The 'far away dimension' no longer functions as synonymous with *país* [country] or *nación* [Nation], but more as an imaginary that becomes a *point de repére* of what is possible/realizable in the ordinary time of the society/community.

TABLE 2.4C. The immigration/emigration condition in the light of the local environment considered as 'amueblamiento'.

emigration discussed up to now. In this sense, then, we have three different possibilities that are presented in Tables 2.4a to 2.4c. When viewed from the environmental perspective, considered from the 'ambiente' point of view, immigration/emigration appears (i) as a *continuum*, (ii) as cultural multiplicity and (iii) as a *condition* of a 'war environment' (see Table 2.4a).

When viewed from the environment perspective, considered as 'scenario', immigration/emigration appears as variable groups of communities situated and organized around *commodities* and values in a financial sense (see Table 2.4b). When regarded from the environment perspective, considered as 'amueblamiento', the immigration/emigration appears as a spatial setup determined by a general sense of *information*, by a standardization in *common values* of things, objects and people. The absence of a traditional topographical dimension is also a part/condition of the experience of perceiving and understanding the future (see Table 2.4c).

To conclude, we should finally consider a general condition of the local space which appears on many occasions, not as *absent* but as *absence*, and it is precisely this 'absence' that characterizes the space itself. It is in connection with this 'absence' that the major local ideas about *time* are paradoxically developed. In this many authors follow the Western/European perspective of G. Deleuze, for whom there is an 'imagination' of time that is produced by breaking the link between movement and time and, in addition, by submitting movement, understood as image, to an *artificial* sense of time (see for example Deleuze, 1985: 354 ff.). That is why in the local domain of 'thinking', the Western/European domain of 'metaphysics' has generated so many reactions and interpretations. And it is not surprising that, given the connection explored by specialists, within Western/European philosophy, between the development of the concept of 'individuality' and the idea of metaphysics (see for example Gracia, 1975, 1984 and 1988; also Miró Quesada, 1963), a number of relations can also be explored locally between the colonial concerns about Western/European metaphysics and the realization of political systems and/or sociological systems where the notion of individuality appears to be relevant. And this is where we can also partially ascribe the sense of 'far away' that runs through the whole sense of local culture.

The transcendental condition of *far away* as a basic condition of thinking and as a fundamental part of the local culture has already been analyzed elsewhere (see for example Canaparo, forthcoming: Chapter 9 and Chapter 10). However, it is relevant to mention now that the 'thinking space' is also established/constituted around this notion of *far away*. The impact and influence that the ideas of Emmanuel Levinas

(1906–1995),[24] for example, have had locally are directly connected to this condition of the *far away*. The case of Enrique Dussel seems the clearest example (see for example Dussel, 1975).

The 'world-view' as a problem of constructivism (see for example Maturana, 1995a and 1995b) and the issue of a 'general encyclopaedia' as a conceptual problem (see for example Darnton, 1986), both constituting the 'universo de discurso' [universe of discourse] as a problem of *languaging* (see for example Mignolo, 1994b), became the main areas of change and questioning in Western/ European traditions, as well as the grounds for the development of the notion of 'present'.

24 The notion of 'ontology' in Levinas, inscribed in the Western/European tradition of the area of metaphysics after Kant, is the main reason for the existence of this relation and connection. This, however, is not something that we propose to develop in this work.

5. Bibliography

Adorno, Rolena. *Guaman Poma. Writing and Resistance in Colonial Perú* (Austin: University of Texas Press, 2000).
———. *New Studies of the Autograph Manuscript of Felipe Guaman Poma de Ayala* (Copenhagen: Museum Tusculanum Press, 2003).
Alberdi, Juan Bautista. *Bases y punto de partida para la organización política de la República Argentina* (Valparaiso: Imprenta del Mercurio, 1852).
Aricó, José. *Marx y América Latina* (Lima: Centro de estudios para el Desarrollo y la Participación, 1980).
Beorlegui, Carlos. *Historia del pensamiento filosófico latinoamericano* (Bilbao: Universidad de Deusto, 2006).
Berryman, Phillip. *Teología de la liberación: Los hechos esenciales en torno al movimiento revolucionario en América Latina* (Mexico: Siglo XXI, 1989).
Bhabha, Homi K. *The Location of Culture* (London: Routledge, 1994).
Blondel, Éric. *Nietzsche, le corps et la culture* (Paris: L'Harmattan, 2006).
Boi, Luciano. *Le probleme mathematique de l'espace* (Berlin: Springer, 1995).
Boi, Luciano et al. (eds.). *1830–1930: A Century of Geometry* (Berlin: Springer, 1992).
Borges, J. L. *Obras completas I* (Buenos Aires: Emecé, 1989).
Bourdieu, P./Sayad, A. *Le déracinement* (Paris: Minuit, 1964).
Brokaw, Galen. 'The Poetics of Khipu Historiography: Felipe Guaman Poma de Ayala's Nueva Corónica and the Relación de los quipucamayos', in *Latin American Research Review*, volume 38, number 3, October 2003, pp. 111–147.
Brotherston, Gordon. *Image of the New World. The American Continent portrayed in Native Texts* (London: Thames and Hudson, 1979).
———. *Book of the Forth World: Reading the Native Americas Through their Literature* (London: Cambridge University Press, 1992).
———. *Painted Books from Mexico. Codices in the United Kingdom collections and the World they Represent* (London: British Museum, 1995).
———. *Footprints Through Time: Mexican Pictorial Manuscripts* (Bloomington: Lilly Library, 1997).
Canaparo, Claudio. *Ciencia y escritura* (Buenos Aires: Zibaldone, 2003).
———. *Muerte y transfiguración de la cultura rioplatense* (Buenos Aires: Zibaldone, 2005).
———. 'La consumación del realismo', in G. Fabry/C. Canaparo, eds., *El enigma de lo real. Las fronteras del realismo en la narrativa del siglo XX* (Bern: Peter Lang, 2007), pp. 199–275.
————. *El imaginario Patagonia. Ensayo acerca de la evolución conceptual del espacio* (forthcoming).
Cañizares-Esguerra, Jorge. *How to Write the History of the New World* (Stanford: Standford University Press, 2001).

Cevallos-Candau, Francisco et al. *Coded Encounters: Writing, Gender, and Ethnicity in Colonial Latin America* (Amherst: University of Massachussetts, 1994).
Congost, Rosa. *Tierras, leyes, historia* (Barcelona: Crítica, 2007).
Cummins, Tom. 'Representation in the Sixteenth Century and the Colonial Image', in Hill Boone, Elizabeth/Mignolo.W. D. (eds.), *Writing Without Words* (Durham: Duke University Press, 1994), pp. 188–219.
Darnton, Robert. *The Business of Enlightenment: Publishing History of the Encyclopédie, 1775–1800* (Cambridge, Mass.: Belknap Press, 1986)
Deleuze, Gilles. *Cinéma 2. L'image-Temps* (Paris: Minuit, 1985).
Derrida, Jacques. 'Introduction' in H. Husserl, *L'origine de la géometrie* (Paris: PUF, 1962).
———. *Marges de la philosophie* (Paris: Minuit, 1972).
———. *L'archéologie du frivole* (Paris: Galilée, 1973).
———. *La carte postale – De Socrates à Freud et au-delà* (Paris: Aubier/Flammarion, 1980).
———. *La filosofía como institución* (Barcelona: Granica, 1984).
———. *Le problème de la genèse dans la phenomenologie de Husserl* (Paris: PUF, 1990).
Derrida, Jacques et al. *Dire l'événement, est-ce possible?* (Paris: L'Harmattan, 2001).
Durán Guardeño, Antonio J. (ed.). *El legado de las matemáticas* (Universidad de Sevilla: Seville, 2000).
Dubnick, Heather. 'Literary Philosophers: Borges, Calvino, Eco', in *MLN*, volume 117, Number 5, December 2002, pp. 1143–1148.
Dussel, Enrique. *Liberación Latinoamericana y Emmanuel Levinas* (Buenos Aires: Bonum, 1975).
Eco, Umberto. *Lector in fabula* (Milan: Bompiani, 1979).
Edwards, Paul N. 'Meteorology as Infrastructural Globalism', in *Osiris*, volume 21, 2006, 229–250.
Feinghold, Mordechai. *Jesuit Science and the Republic of Letters* (Cambridge, Mass.: MIT Press, 2003).
Foerster, Heinz von. *Cybernetics of Cybernetics* (Minneapolis: Future Systems Inc., 1995).
Foucault, Michel. *Histoire de la sexualité. 3 Le souci de soi* (Paris: Gallimard, 1984).
———. *Sécurité, territoire, population* (Paris: Gallimard/Seuil, 2004). [2004a]
———. *Naissance de la biopolitique* (Paris: Gallimard/Seuil, 2004). [2004b]
Galeano, Eduardo. *Open Veins of Latin America: Five Centuries of the Pillage of a Continent* (New York: Monthly Review Press, 1973).
———. *Patas arriba. La escuela del mundo al revés* (Mexico: Siglo XXI, 1998).
Gallison, Peter/Thompson, Emily (eds.). *The Architecture of Science* (Cambridge, Mass.: MIT Press, 1999).
García Canclini, Néstor. *Culturas híbridas. Estrategias para entrar y salir de la modernidad* (Buenos Aires: Paidós, 1999).
———. *Consumers and Citizens: Globalization and Multicultural Conflicts* (Minneapolis: University of Minnesota Press, 2001).

Geertz, Clifford. *The Interpretation of Cultures* (Basic Books: New York, 1977).
———. *Local Knowledge: Further Essays in Interpretive Anthropology* (Basic Books: New York, 1985).
———. *Works and Lives: The Anthropologist as Author* (Stanford: Stanford University Press, 1997).
Glasersfeld, Ernst von. *Radical Constructivist. A Way of Knowing and Learning* (London: The Falmer Press, 1995).
González Casanova, Pablo, (contrib.). *América Latina. Historia de medio siglo* [volume 2] (Mexico: Siglo XXI, 1981).
———. *El estado en América Latina: Teoría y práctica* (Mexico: Siglo XXI, 1990).
Goodman, Nelson. *Fact, Fiction and Forecast* (Cambridge, Mass.: Harvard University Press, 1983).
Gracia, Jorge J. E. 'Importance of the History of Ideas in Latin America: Zea's *Positivism in Mexico*', in *Journal of the History of Ideas*, volume 31, number 1, pp. 177–184.
———. *Introduction to the Problem of Individuation in the Early Middle Ages* (Washington DC: The Catholic University of America Press, 1984).
———. *Individuality: An Essay on the Foundations of Metaphysics* (Albany: State University of New York, 1988).
———. *Philosophy and Its History: Issues in Philosophical Historiography*, (Albany: SUNY Press, 1992).
———. *La filosofía y su historia* (Mexico: UNAM, 1998).
———. 'Borges' "Pierre Menard": Philosophy or Literature?', in *The Journal of Aesthetics and Art Criticism*, volume 1, number 1, Winter 2001, pp. 45–57.
Griffin, Nicholas. 'Review of Torretti', in *The Philosophical Quarterly*, vol. 31, num. 125, 1981, pp. 374–376.
Haraway, Donna J. *Simians, Cyborgs, and Women. The Reinvention of Nature* (London: Free Association Books, 1991).
Harrison, Stephan, et al. (eds.). *Patterned Ground. Entanglements of Nature and Culture* (London: Reaktion, 2004).
Herman, Edward S./Chomsky, Noam. *Manufacturing Consent. The Political Economy of the Mass Media* (New York: Phanteon Books, 1988).
Hilbert, D. *Grundlagen der Geometrie* (Stuggart: Teubner, 1968). Originally published in 1899.
———. *Foundations of Geometry* (Chicago: Open Court, 1971).
Hill Boone, Elizabeth. 'Introduction: Writing and Recording Knowledge', in Hill Boone, Elizabeth/Mignolo.W. D. (eds.), *Writing Without Words* (Durham: Duke University Press, 1994), pp. 3–26.
Huff, Toby. 'Rethinking Civilizational Analysis', in *The Canadian Journal of Sociology*, volume 31, number 3, Summer 2006, pp. 375–379.
Jameson, Frederic/Miyoshi, Masao. *The Cultures of Globalization* (Durham: Duke University Press, 1998).
Justo López, Mario. *Introducción a los Estudios Políticos* (Buenos Aires: Depalma, 1982). [2 volumes]

Klein, Felix. *Vergleichende Betrachtungen über neuere geometrische Forschungen* (Erlangen: A. Duchert, 1872).
———. *Elementary Mathematics from an Advanced Standpoint: Geometry* (New York: Dover, 1984).
Kusch, Rodolfo. *América profunda* (Buenos Aires: Hachette, 1963).
———. *El pensamiento indígena y popular en América Latina* (Buenos Aires: Hachette, 1970).
———. *Obras Completas* (Rosario: Fundación Ross, 2000). [2 volumes]
Larrain, Jorge. *Identity and Modernity in Latin America* (London: Polity, 2000).
Latour, Bruno. *Reassembling the Social. An Introduction to Actor-Network-Theory* (Oxford: Oxford University Press, 2005).
Latour, B./Weibel, P. (eds.). *Making Things Public. Atmospheres of Democracy* (Cambridge, Mass.: MIT Press/ZKM, 2005).
Lebeau, André. *L'engrenage de la technique. Essai sur une menace planetaire* (Paris: Gallimard, 2005).
Lechner, Norbert (comp.). *Estado y política en América Latina* (Mexico: Siglo XXI, 1981).
Leibsohn, Dana. 'Primers for Memory: Cartographic Histories and Nahua identity', in Hill Boone, Elizabeth/Mignolo, W. D. (eds.), *Writing Without Words* (Durham: Duke University Press, 1994, pp. 161–187).
Lévy, Jacques. *Le tournant géographique* (Paris: Belin, 1999).
Livingstone, David N. *Putting Science in Its Place. Geographies of Scientific Knowledge* (Chicago: Chicago University Press, 2003).
Lobachesvki, N. I. 'Géometrie imaginaire', in *Journal für die reine und angewandte Mathematik*, vol. 17, 1837, pp. 295–320.
Lugones, Leopoldo. *El tamaño del espacio (Ensayo de Psicología Matemática)* (Buenos Aires: El Ateneo, 1921).
Magnani, L. *Philosophy and Geometry: Theoretical and Historical Issues* (Dordrecht: Kluwer, 2001).
Marcus, Joyce. *Mesoamerican Writing Systems: Propaganda, Myth, and History in Four Ancient Civilizations* (Princeton: Princeton University Press, 1992).
Maturana, Humberto. *La realidad: ¿objetiva o construida? I. Fundamentos biológicos la realidad* (Barcelona: Anthropos, 1995). [1995a]
———. *La realidad: ¿objetiva o construida? II. Fundamentos biológicos del conocimiento* (Barcelona: Anthropos, 1995). [1995b]
Maturana, Humberto/Varela, Francisco. *Autopoiesis and Cognition* (Dordrecht: Kluwer, 1980).
Mazlish, B./Buultjens, R. (eds.). *Conceptualizing Global History* (Boulder: Westview Press, 1993).
Medin, Tzvi. *Leopoldo Zea. Ideología, historia y filosofía de América Latina* (Mexico: UNAM, 1983).
Mignolo, Walter. 'Signs and Their Transmission: The Question of the Book in the New World', in Hill Boone, Elizabeth/Mignolo.W. D. (eds.), *Writing Without Words* (Durham: Duke University Press, 1994, pp. 220–270). [1994a]

———. 'Afterword: Writing and Recorded Knowledge in Colonial and Postcolonial Situations', in Hill Boone, Elizabeth/Mignolo.W. D. (eds.), *Writing Without Words* (Durham: Duke University Press, 1994, pp. 292–310). [1994b]
———. *The Darker Side of the Renaissance. Literacy, Territoriality and Colonization* (Ann Arbor: Michigan University Press, 1995).
———. *Local Histories/Global Designs. Coloniality, Subaltern Knowledge and Border Thinking* (Princeton: Princeton University Press, 2000).
———. *Historias locales/diseños globales. Colonialidad, conocimientos subalternos y pensamiento fronterizo* (Madrid: Akal, 2003).
———. *The Idea of Latin America* (London: Blackwell, 2005).
Miller, Clark. A./Edwards, Paul N. (eds.). *Changing the Atmosphere: Expert Knowledge in Environmental Governance* (Cambridge, Mass.: MIT Press, 2001).
Miró Quesada, F. 'The Impact of Metaphysics on Latin American Ideology', in *Journal of the History of Ideas*, volume 24, 1963, pp. 539–552.
Murena, Héctor A. *Homo Atomicus* (Buenos Aires: Sur, 1961).
Nicholson, Max. *La revolution de l'environment* (Paris: Gallimard, 1973).
Nohlen, Dieter. 'Ciencia Política en América Latina', in D. Nohlen (ed.), *Diccionario de Ciencia Política* (Mexico: Porrúa, 2006). [2 volumes].
Nozick, Robert. *Anarchy, State and Utopia* (London: Blackwell, 1978).
———. *Invariances: The Structure of the Objective World* (Cambridge, Mass.: Harvard University Press, 2003).
Orloff, Ann/Clemens, Elizabeth. *Remaking Modernity: Politics, History, and Sociology* (Durham: Duke University Press, 2005).
Ortiz, Eduardo. 'A Convergence of Interests: Einstein's Visit to Argentina in 1925', in *Ibero-Amerikanisches Archiv*, Jahrgang 21, 1–2, 1995, pp. 67–126.
———. 'The Transmission of Science from Europe to Argentina and its Impact on Literature: from Lugones to Borges', in E. Fishburn, ed., *Borges and Europe Revisited* (London: ILAS, 1998), pp. 108–123.
———. 'On the Transition from Realism to the Fantastic in the Argentina of the 1870s: Holmberg and the Córdoba Six', in E. Fishburn/E. Ortiz, eds., *Science and the Creative Imagination in Latin America* (London: Institute for the Study of the Americas, 2005), pp. 59–85.
Osborn, Roger. 'Some Historic and Philosophic Aspects of Geometry', in *Mathematics Magazine*, vol. 24, num. 2, 1950, pp. 77–82.
Pachacuti Yamqui, Joan de Santa Cruz. *Relación de antigüedades deste reyno del Piru* (Cusco: Centro Bartolomé de las Casas, 1993).
Perlongher, Néstor. *Papeles insumisos* (Buenos Aires: Santiago Arcos, 2004).
Piaget, Jean. *La construction du réel chez l'enfant* (Neuchâtel: Delachaux and Niestlé, 1937).
———. *Biologie et Connaissance* (Gallimard: Paris, 1967).
Pratt, Mary Louise. *Imperial Eyes. Travel Writing and Transculturation* (London: Routledge, 1992).

Pohl, John M. D. 'Mexican Codices, Maps, and Lienzos as Social Contracts', in Hill Boone, Elizabeth/Mignolo.W. D. (eds.), *Writing Without Words* (Durham: Duke University Press, 1994), pp. 137–160.

Popkin, Richard. 'Review on Gaukroger', in *The British Journal for the Philosophy of Science*, vol. 34, num. 2, 1983, pp. 182–185.

Riemann, B. 'Über die Hypothesen, welche der geometrie zugrunde liegen', in *Abhandlungen der Kgl. Gesellschaft der Wissenschaften zu Göttingen*, vol. 13, 1867, pp. 133–152.

Rosenfeld, B. A. *A History of Non-Euclidean Geometry: Evolution of the Concept of a geometric Space* (New York: Springer, 1988).

Ross, Andrew. *Strange Weather: Culture, Science and Technology in the Age of Limits* (London: Verso, 1991).

Rushdie, Salman. *Imaginary Homelands* (London: Granta, 1991).

Russell, Bertrand. *An Essay on the Foundations of Geometry* (Cambridge: Cambridge University Press, 1897).

Sabato, Ernesto. *Hombres y engranajes* (Buenos Aires: Emecé, 1951).

Sahlins, Marshall. *Culture and Practical Reason* (Chicago: Chicago University Press, 1978).

———. *Islands of History* (Chicago: Chicago University Press, 1987).

———. *Crisis in Western Sociological Understanding* (London: Berg, 1990)

———. *How 'Natives' Think: About Captain Cook, for Example* (Chicago: Chicago University Press, 1995).

Said, Edward. *Culture and Imperialism* (London: Vintage, 1994).

———. *Out of Place: a Memoir* (London: Granta, 2000).

Salazar Bondy, A. *La cultura de la dependencia* (Lima: Instituto de Estudios Peruanos, 1966).

Sayad, Abdelmalek. *L'immigration ou les paradoxes de l'alterité. 1. L'illusion du provisoire* (Paris: Raison d'Agir, 2006). [2006a]

———. *L'immigration ou les paradoxes de l'alterité. 2. Les enfants illégitimes* (Paris: Raison d'Agir, 2006). [2006b]

Schäfer, Wolf. *Finalization in Science. The Social Orientation of Scientific Progress* (New York: Elseiver, 1983).

———. 'Global History: Historiographical Feasibility and Environmental Reality', in Mazlish, B./Buultjens, R. eds., *Conceptualizing Global History* (Boulder: Westview Press, 1993), pp. 47–69.

Schneider, Arnd. *Futures Lost: Identity and Nostalgia among Italian Immigrants in Argentina* (Bern: Peter Lang, 2000).

Sennett, Richard. *The Corrosion of Character* (New York: Norton, 1998).

Shenann, Douglas H. *Evolution and the Nature of Reality* (Victoria: Trafford Publishing, 2006).

Soja, Edward. *Postmodern Geographies. The Reassertion of Space in Critical Social Theory* (London: Verso, 1989).

Spitta, Silvia. *Between Two Waters. Narratives of Transculturalization in Latin America* (Austin: Texas and AM University Press, 1995).

Thorp, Rosemary. *Progress, Poverty and Exclusion. An Economic History of Latin America in the 20th Century* (Washington: IDB, 1998).
Todorov, Tzvetan. *La conquista de América. La cuestión del otro* (Mexico: Siglo XXI, 1987).
Torres García, Joaquín. *La tradición del hombre abstracto (Doctrina constructivista)* (Montevideo: edición de autor, 1938).
———. *La recuperación del objeto* (Montevideo: edición anónima, 1948).
Tuner, Charles. *Modernity and Politics in the Work of Max Weber* (London: Routledge, 1992).
Whyte, Jennifer. *Virtual Reality and the Built Environment* (New York: Elsevier, 2002).
Zamora, Margarita. *Reading Columbus* (Berkeley: University of California Press, 1993).
———. *Language, Authority, and Indigenous History in the 'Comentarios Reales de Los Incas'* (New York: Cambridge University Press, 2005).
Zemelman, Hugo. *De la historia a la política. La experiencia de América Latina* (Mexico: Siglo XXI, 1989).

Chapter 3
Science and Empire

> Different interpretations presuppose a common and shared principle of knowledge and of the rules of the game, while different perspectives presuppose that the principles of knowledges and the rules of the game are geo-historically located in the structure of power of the modern colonial world. To show how this works, we need something such as 'dependency theory' for the epistemological domain.
>
> Walter D. Mignolo
> *The Idea of Latin America*

We have been exploring the origins and conditions of local 'thinking'. We have also explored the notion of 'spatial thinking' as a local condition of speculation and as a framework for understanding the evolution of culture and ideas. Moreover, as we have already hypothesized, the relationship between what we understand as 'local' and as 'spatial thinking', is direct and difficult to predict as both elements define each other. Furthermore, the connection between the creation of a local space and the determination of a particular notion and perspective of knowledge is also direct. In Chapter 4 we will specifically explore the local notion of knowledge. However, before we can begin that discussion, we need to determine the ground upon which knowledge has been understood and conceived in the peripheral and colonial circumstances of the local 'thinking space'.

Why the local notion of knowledge has been closely defined in relation with a particular idea of 'science' is something that we have mentioned before. However such an important aspect and development deserves extended consideration and that is what will be developed within the present chapter. And it is because of this condition and context that the already lengthy discussion about 'science and empire' between the critics and local authors needs to be addressed. It is not only a matter of discussing a particular set of concepts but also of exploring (i) why these issues became so important locally and (ii) what cultural and intellectual consequences such historiographical perspectives have locally. Therefore we will divide the present chapter into seven parts:

1. We will examine the meaning of historiography in the context and terms of the 'historiographical connection', as described in Chapter 2;
2. We will consider the idea of 'geo-epistemology' in relation to the subject;
3. Following on from the previous point, we will explore the evolution of the ideas in such context.
4. We will study a notion of translation as a local specific cultural characteristic;
5. The relationship between empire and science will be examined from the spatial point of view;
6. We will demonstrate/exhibit the main concepts considered by the authors and why (in these circumstances).
7. We will show the current state of all theses aspects in the present global and visual culture we are living in.

1. A Historiographical Construction

The study of empires is not a recent academic subject. What is more recent is the study of empires in relation to the transmission of knowledge or, more specifically, in relation to the imposition/introduction of a particular notion of *science*. The study of the notion of empire (the conceptual issue) becomes entangled with the consideration of the beginning of the Great Expeditions period, when Spain was still a relevant, if rather decadent, empire in the Western World (the historiographical issue). The Malaspina expedition of 1789–1794 is probably the best recent example. Even more contemporary is the development of science and Empire as a subject within the university environment, an event which corresponds to the post-colonial historical revisions produced after the Second World War. What is important to consider here is the fact that the argument of science and Empire, such as it is (the conceptual dimension), cannot be separated from the historiographical context from which it emerged (the historiographical dimension). Moreover, the expansion of university populations and the diffusion of academic works and analyses that are connected with the establishment of English as the *lingua franca*, can be considered as the ultimate period in the scientific domination of the world through the establishment of standards and parameters of writing.

The consideration and analysis of the relationship between science and Empire as a *historiographical phenomenon* mainly refers to the British Empire because it was British-based publications that first introduced the argument extensively (see for example Greene, 1961; Headrick, 1981; Harrison, 2005; and Ryan, 1997). This was/is the situation even after considering the fact that, as we know, Spain and Portugal's empires were far more extended in historical terms, and more expanded in territorial terms (Cañizares-Esquerra, 2005). Although, chronologically speaking, the study of the Spanish Empire and domains should lead to the study of 'colonial science', the case has been rather the opposite. It was, on the contrary, the development of studies of the British Empire or, indeed, the discussion by Anglo-Saxon authors of empires other than the British (see for example Needham, 1954–1986), that established the historiographical ground for the analysis of the relationship between 'science and Empire' (see for example Bayly, 1989 and Hall, 2000). Cañizares-Esquerra (2005) argues that studies of the Spanish Empire refer mostly to the sixteenth and eighteenth centuries, and marginalize the remaining periods, including the nineteenth and twentieth centuries. In fact, the most comprehensive works of the last fifty years regarding the relationship between science and empire are to be found within 'Anglo-Saxon' publications. I am referring, for example, to Roy MacLeod's 'Nature and Empire: Science and the Colonial Enterprise' published by *Osiris* in 2000 and to Londa Schiebinger's edited section of the journal *Isis*, under the title of 'Colonial Science', in 2005.

Another interesting aspect concerns the incorporation of analyses and writing connected to the relationship between science and empire within the Spanish domain which was actually carried out within the historiographical context set up by Anglo-Saxon analysts and academics. Our hypothesis is that this happened because it is impossible to separate the idea of science and empire from the academic context in which it has been developed since 1850. The construction of the notion of 'British Empire', given its historically orientated nature, led to the establishment of the connection between science and empire. The notion of a certain

'Imperial Science', for example, was already present in the pages of journals like *Nature* in the 1890s (see for example Canaparo, 2003; and, more generally, see Headrick, 1981; Drayton, 2000; and MacKenzie, 1988). Some earlier analyses were also provided, with regard to the Spanish Empire during the seventeenth and eighteenth centuries, but this was nothing compared to the enormous bibliography produced by Anglo-Saxon authors during the twentieth century. These earlier British works were historically oriented and did not have the *critical* approach with which we associate the subject today. Focusing less on establishing just how the Empire imposed an idea and/or practice of science, the earlier works were concerned more with the construction and expansion of ideas and their development within the colonies (see for example MacLeod, 2000). However, more importantly, (i) the philosophical notion of naturalism (see for example Arnold, 1996; MacKenzie, 1988 and 1990; Young, 1958), (ii) the idea of experiment (see for example Latour, 1985) and (iii) the notion of modernity expressed in terms of scientific progress (see for example Macleod, 2000; Drayton, 2000) made it almost impossible for the scientific practitioner to construct any sociological theory of science in the *critical* sense in which we understand it today. The works of Andrés Galera Gómez and Juan Pimentel with regard to Malaspina's expedition (see Galera Gómez, 1988 and Pimentel, 1998) as well as those relating to Latin America by authors such as Charles Darwin (1809–1882) (see for example Darwin, 1989; Greene 1961; and Young, 1985) and Alexander von Humboldt (1769–1859) (see for example Humboldt, 1995 and Minguet, 1969) are seminal examples in this respect.

Despite all the differences and changes of the various approaches, as briefly indicated above, an *argumentative line* can be constructed between the civil servants and the first scientific practitioners who approached the notion of scientific diffusion using the imperial structure and the present Anglo-Saxon University system which considers the *subjects* of science and empire under the more contemporary light of 'Science Studies' or 'Technological Studies' (see for example Osborne, 2000). To summarize: the argument of science and empire in its present form is a result of the expansion and development of the Anglo-Saxon University system and publishing market. This situation is confirmed by specialists like Roy Macleod who situates the study of the connection between science and empire – as 'area study' – as beginning during the 1960s and 1970s (see Macleod, 2000). During the years of German-based development of certain sciences, from the eighteenth to the twentieth centuries (see for example Pyenson 1985 and 1993), this notion of empire only arose in political terms and not in *scientific* terms. This can be considered proof of the previous argument. Moreover, in terms of science and *techniques*, Western/European culture was considered, for a great part of the nineteenth century, in terms of *civilization* (see for example Burckhardt, 1914). It is only during the twentieth century that, from a philosophical or scientific point of view, the notion of Western/European culture was presented in relation to different empires. Even for authors like Immanuel Kant the association between 'science' and knowledge was less justified in the particular context of cultural progress, a Eurocentric view (see Mignolo, 2005), than by reference to the idea of *connaissance de la nature* (see for example Foucault, 1970). In fact, it was precisely from this notion of *civilization* that the idea of something called 'Western culture' was created. Oswald Spengler's (1880–1936) classic work of *The Decline of the West* (see Spengler 1926 and 1928) can be understood in this sense (see for example Fischer, 1989). The fact that the entire de-colonization period was, from the political/institutional point of view, only formally over at the end of 1960s, is certainly a further factor to explain why it was only then that the first studies about a sort of 'colonial/imperial science' emerged. The already classic and widely-quoted work of George Basalla 'The Spread of Western Science' (1967) is indeed a perfect example. We can also mention the works of K. Mendelssohn (1976) and A. Alam (1977).

It is interesting to see how a historiographical account of the connection between science and the British Empire can be considered as a model of the relationship between empire and science as such. This is less to do with a particular author's claims than with the fact that the English language functions as *lingua franca* and that Anglo-Saxon universities are considered as the *natural* location for Western/European contemporary knowledge. Therefore, either because the subject refers specifically to the British Empire or because the author deploys an approach within the environment of Anglo-Saxon universities, the result is always a *performative* (see for example Andersson, 1975; Searle, 1989; Parker/ Kosofsky Sedgwick, 1995) condition of the

British Empire. Mark Harrison's article 'Science and the British Empire' (2005) is probably one of the most recent examples whereby the specific claims of the article are superseded by the historiographical context in which it is situated. This is certainly extremely pertinent with respect to the argument of 'science and empire' as it always works in this way: the descriptive level of commentary and analysis does not match the epistemological grounds upon which it is located.

It is not only the facts and arguments considered under the category of 'British Empire' that make the above-mentioned approach *performative*; it is also the way in which it is deployed. The fact that the British Empire was spread out widely across the planet, making it an unquestionable model for imperial analysis, cannot be separated from the fact that its analysis is referred to by the Anglo-Saxon academic and scientific community with a vast bibliography that is published in English and mostly by British and American publishers. Again, Harrison's article (2005) remains a good example: in this respect, the assumption that in England 'the unfolding of empire coincided with the scientific revolution' (Macleod, 2000) is not a minor claim. This situation allows us to understand why many authors associate the development of science with the progress of the 'civilizing mission'.

The argument of science and empire, such as it is, is not only a *historiographical creation*, it is also something that in terms of knowledge cannot be detached from the *location* imposed/established by the development of imperialism itself. If, in these conditions of knowledge, it is paradoxically almost impossible to conceive 'modernity' without colonialism (see Mignolo, 2005), then, equally, 'science' in its modern meaning is always 'colonial science'. Because it involves a central/peripheral relation, it assumes a degree of diffusion and progression, and, most importantly, because it implicitly deploys a methodological approach that is consistent with the creation and development of an Imperial State. Of course, this introduces more of a conceptual and epistemological problem (see for example Deleuze/Guattari, 1980 and 1991) and less of a political or ideological issue in the sense popularized by Toni Negri and Michael Hard in their best-selling book *Empire* (2000). However, contrary to what some authors might think (see for example Cardoso and Faletto, 1979), the fact that science, associated in Europe with the idea of knowledge since Kant's writings (see for example Kant, 1981), always functions in colonial/imperial terms does not facilitate the identification of a particular 'matrix' or scheme of analysis. The connection between science and empire remains very complex and generalizations are difficult to sustain.

In the same way that 'modernity' is not the *historical evolution* ('superación histórica') of colonialism but, on the contrary, the consequence of it (see Mignolo, 2005), the imposition of a paradigm of knowledge – universities, vehicular language, etc. – is not only a way to better understand the relationship between science and empire but also to reproduce it. The connection between science and empire is less about historical consideration or information than about the way in which, through historiographical narrations, we not only construct the past but also establish a conceptual system to explain it. It is interesting to note how, as the argument of colonial/imperial science became more complex and varied, the community of academic (and university) authors that deal with the subject became more galvanized and standardized. Following the ideas of Slavoj Žižek, for example, about contemporary culture, this reverse situation/condition can also be considered as the relationship between what is 'real' and what is 'symbolic' within the present capitalist environment: the more symbolic our understanding becomes, the greater the demand we have to establish a notion of what is 'real' in relation to it (see for example Žižek, 2004). This situation/condition can also be understood in light of the paradoxes exposed by Jean-François Lyotard (1924–1998) in *Le differend* in order to characterize our present sense of thinking. For Lyotard the category of 'modernity' needs to be interpreted less as a set of clear and established rules and more as a result of non-resolved issues, an interpretation that sees the thought of modernity – or 'Modern Thought' – as a continuation of the notion of knowledge introduced by the empires since the fifteenth century (see for example Lyotard, 1988).

2. Empire and Geo-epistemology

More than a concept, empire(s) is a notion whose study and definition cannot be separated from the historical consideration of its evolution and, therefore, from the historiographical development of the notion itself (i) as a historical category, (ii) as a political system or (iii) as a form of domination. Whatever aspect is analysed in relation to this notion of empire(s), this particular condition of evolution and historiographical development must be taken into account, firstly as a historical category (temporal concept, historicity), secondly, as a political structure or administration and, finally, as a socio-cultural form of domination. From a historiographical perspective, empire re-defines the past as a form of domination that can be interpreted, according to the period, as a political system or, more specifically, as an historical category. Furthermore, the analysis and definition of the notion of empire usually follows the three above-mentioned instances. In other words, within a practical use and implementation of the definition of empire, three perspectives can be considered: (i) as a political argument and/or system, (ii) as an ideological dimension, and (iii) as a technological/scientific point of view.

As a political argument, the idea of empire has been analysed as a product of the State, Government or Nation-orientated administration. This is the classical approach to empire when it is considered within a historical context and, consequently, as part of a particular State-orientated strategy. It is mostly in this sense that authors refer to the 'Roman Empire', 'Spanish Empire', etc. From this perspective, empire can appear as the main institution that allows a panoramic view of the State, the administration, etc., and that is thus comprehensive of almost every domain. Extreme versions of these standpoints, conceived the idea of empire as the ultimate political *logos* (see for example Huntington, 2002). However, empire can also be understood here, in a more specific way, as the political system itself.

Conceived as a product of a particular paradigm, the empire is connected with a system of beliefs and/or a set of values that are implemented at the social level. The most common way of describing this conception is the ideological approach, that is, the idea that there is a set of *values* that are coherent, conform to a sort of *scheme*, and can be identified through the *analysis of social actors*. This approach has dominated the idea of empire for the whole of the twentieth century, especially for those authors for whom the analysis of history or society was a manifestation of their political beliefs (see for example Galeano, 1973). The notorious principle of *politique d'engagement* has its roots in this approach, as does the development of the notion of *imperialism*. Since the beginning of the twentieth century, the evolution of the category of empire, in terms of ideology, has followed a sort of *escape to invisibility*: rather than grounding the effectivity and efficacy of empire in monuments or in wars for possession of land, empire has been transformed into imperialism, and its roots have been transformed into more subtle forms of domination.

However, there is a more recent approach, somehow derived from the previous ideological perspective. This conceives empire less as a political or physical entity and more as being connected to non-visible elements, that is, in relation to scientific and technological elements. The most powerful of these perspectives, developed at the end of the twentieth century, seems to consider science and technology as *devices* of imperialism. In short, according to this perspective, the combination of science and technology, its use and development, together within an invisible sense of reality, have generated a reproduction of imperial structures of organizing space, understanding it and 'naming' it in peripheral communities or societies. This perspective, in a strict sense, is the factor that generates the increasing analysis of the relationship between science and imperialism and, moreover, is the perspective that establishes it as a *subject*.

Finally, there is one last argument that has been developed recently and which considers empire as a *substitute notion*, that is, a category that is relevant only if placed in its geo-epistemological context (see for example Mignolo, 2000 and 2005). This approach tries to establish a link between spatial location ('geography' = physical and imaginary) and conceptual evolution ('knowledge') (see for example Deleuze/ Guattari, 1980 and 1991), and to discover how each particular location develops a specific way of conceiving knowledge and therefore a sort of 'independent epistemology' in which every language (and conceptual domain) is grounded. It is in the light of this later approach that we would propose to explore the relationship between science and empire.

With the historical approach we face the limits of a factually and chronologically outdated analysis. With the ideological approach we have the problem of universal conceptions applicable to every single case and also the idea of regularities, that is to say, recurrent cycles and periods. With the 'scientific' approach we confront the question of knowledge and of by whom and how it is defined. With the geo-epistemic approach there is a double attempt: (i) to try to overcome the limitations of the previous interpretations of empire, and (ii) to try to construct a more accurate epistemological dimension while linking it more closely with a local space rather than with a global-imperial conception.

Given these indicated conditions and characteristics, the analysis of the relationship between an idea of science and a notion of empire becomes equivalent to the analysis of the development and evolution of these three instances of the notion of empire. Why? Because the idea of science as a 'modern' entity cannot be separated from the establishment of the political system that created the European States, during the nineteenth century, in their present shape. There is a direct conceptual link between conceiving a notion of 'modernity', defining the role of the State (imperialism, colonialism) and establishing an idea of science as a *natural* knowledge (see for example Jardine et al., 1996; MacKenzie, 1990 and 1998; and Miller et al., 1996; see also point 6 later in this chapter).

This is relevant because, just as the idea of empire (and imperialism) cannot be considered separately from the notion of modern State and the meaning of modernity, so the modern idea of science (and scientificism) cannot be separated from (i) the university system established in Europe in the nineteenth century and later in the US during the twentieth century and, also, from (ii) the hierarchical bureaucracies deriving from the evolution of the modern State, and (iii) the development and generalization of the market as a public domain. It is almost impossible nowadays to think in terms of science without considering this entanglement and, therefore, to assume that the conceptual evolution of the idea of empire(s) goes together with the conceptual evolution of the notion of modern science.

3. The Evolution of the Ideas

3.1 The Historiographical Foundation of Empires

The relationship between science and empire analysed from the perspectives discussed above can also be described as an evolving development process, not always chronologically organized, which follows five main periods. These can be summarized as follows. (i) The *historical period.* The relationship between science and empire is founded on historical events, periods or schemes. (ii) The '*war conflict*' *period.* The connection between science and empire is based on the particularities and crisis conditions of a war environment. (iii) The '*political*' *period.* It is the particular political system and/or structure that creates the favorable conditions for the entanglement between an idea of science and a notion of empire. (iv) The *specific scientific area and/or period.* This is when it is understood that a particular period of a discipline – for example, history of medicine during nineteenth century – or a scientific area – for example, quantum mechanics – allows for a clear understanding/comprehension and better interpretation of the relationship between science and empire. (v) The *post-colonial and 'Science Studies' period.* This is the name used to indicate development since the 1980s, referring not only to the characteristics of the relationship between both notions, but also to the fact that, as concepts, science and empire became entangled and became part of the same *epistémé*, to use Michel Foucault's expression (see Foucault, 1970).

The relationship between science and empire, in its modern Western/European meaning, that is, from a historiographical point of view, at least since the seventeenth century, can be analysed and presented according to the five periods indicated above. However, these periods are also used to cover the *subject* of empire even if it was not yet considered as such: there is an important difference between the *notion* of Empire and the *concept* of it. This distinction – and these two forms – constitute what we can call the *epistemological ground* of the meaning of empire. As a *notion*, empire can be tracked anywhere in history and chronologies, while, as a *concept*, empire is a recent phenomenon. As a notion, empire has many forms, and it was not always acknowledged as such: it is an *argument* within history. But as a concept, empire functions at a *historiographical level*, that is, empire as a concept allows us to re-write both the past and the future. The five aforementioned periods are not necessarily successive, or chronologically established; they operate more at random and in alternate ways. The definitive entanglement between science and empire in Western/European domains occurs when, towards the end of the nineteenth century, an *imaginary dimension* and a *material dimension* of the notion of empire began to be considered as *concepts* (see point 3.3). It was at that moment that (i) science also became associated with knowledge, and (ii) empire became associated with the modern State, capitalism and liberal democracy.

Nowadays, the notion of empire, that is empire as a simple *historical argument*, is associated with information and does not have much theoretical relevance for authors dealing with it as a specific subject (see for example Benton, 1994; and Baber, 1996). Furthermore, in some cases this notion of empire is associated with a certain administrative and bureaucratic point of view: a government or State-orientated perspective (see for example Hefferman, 1994; Pimentel, 2000; and Gascoigne, 1999). In fact, the recent works dedicated to the relationship between science and empire are clearly a product of, among other things, the *historiographical* consideration of empire as a concept (see for example Baber, 1996). The idea, for example, of the existence of something called 'colonial science' is a clear indication of this situation (see for example Schiebinger, 2005). Paradoxically, beginning with Friedrich Nietzsche and ending with Sigmund Freud, one can observe, on one hand, an agreement with regard to the decadence of European culture and, on the other hand, academics celebrating a 'triumphal' notion of European science and State. In this respect, it is not surprising that many authors have recently postulated as non-compatible the rise of a technological and scientific culture opposed another humanitarian and artistic culture that is falling (see for example Steiner, 1963 and 1971; also the debate about the 'methodological unit' of science in Snow, 1959 and 1964; Canaparo, 2003).

3.2 Science as Knowledge

Following the previous point we can establish that the relationship between science and Empire, despite the fact that they started to be mentioned as related arguments during the nineteenth century, only emerged as an analytical area/domain when Empire as a concept was already in place. Equally important, they also emerged at a time when 'science' was considered as pure knowledge in an empirical and a philosophical sense. It was only towards the end of the nineteenth century and the beginning of the twentieth century, when science started to be considered as the most advanced and 'perfect' form of knowledge and philosophy, that from the analytical point of view it became coherent to consider the relationship between science and empire as natural and necessary. In this sense, the emergence of the analytical relationship between science and empire cannot be separated from the development of the philosophy and history of science, as in fact it cannot be separated from the development of a 'scientific culture' aimed at dominating our contemporary *social imaginary*, as is very clear in a number of works by authors such as Bruno Latour (see for example Latour/Weibel, 2005).

Therefore, the idea of science and empire is directly connected with a notion of *knowledge*. It was this notion, in philosophical and European terms, that facilitated the introduction in the colonies of the *universalization of reality* through the idea of 'naturalism', 'rational language' and 'scientificity'. This was a completely new and radical cultural phenomenon since the disappearance of the idea of God as the 'Universal Master' (see for example Wallerstein, 1974). In practical terms, this universalization was implemented by (i) an idea of education, (ii) progress based on institutional achievements, and (iii) the enlightenment conception of communication, a set and ensemble that prevails even today. In order to overcome this situation and to avoid the simplification of standardization ('globalization'), a number of authors introduced the category of *translation* as a fundamental and basic concept in the study of philosophical, scientific and cultural evolution (see for example Latour, 1987 and Montgomery, 2000). Consequently, the idea of science and knowledge as a corpus and/or narratives that are *translated* into another culture represented an approach that somehow acknowledged the questions raised by the notion of *location of knowledge* previously indicated. Thus, this *translation* features two immediate dimensions: the material and the imaginary.

3.3 The Material and the Imaginary Dimensions

The first dimension, the *material* dimension, is connected with constructions (architecture), instruments (laboratory, experiments) and objects (writing, museums/'museum-ing', archives/archiving). Constructions have been created locally with the help of foreign designers or based on the local *interpretations* of what was done elsewhere (see for example Arnold, 1988; Macleod, 1987; Sheets-Pyenson, 1988). As for instruments and objects, they were first *imported* and then copied/reproduced, including local *adaptations* or, when the object was *localized* within the colony space, it was their meaning, taxonomy and interpretation that were imported. The second dimension, the *imaginary* dimension, is connected with writing, images and theories (see for example Ryan, 1997; Gregory, 1994; and Pratt, 1992). This dimension was much more interactive, and mutual exchanges, between the present and the past, go back to the 'discovery' of America itself. This discovery can be considered in this sense as the primitive origin of *imperialism* in its Western/European and modern form (see for example Mignolo, 2005; also point 6 below). However, the multiple uses and mutations of theories and images did not substantially modify, for example, the present direction and/or composition of the structure of the scientific community, or the editorial market. The understanding of being in a peripheral location and having a 'secondary grading' for local developments not only became commonplace, it also became a way of thinking or, more precisely, a pre-condition of the understanding. Therefore, the more the components of this dimension proliferated, the more the understanding of knowledge itself became alien to the locals.

3.4 The Levels of Engagement

The above-mentioned *translation* was operating, contemporaneously with these two previously mentioned dimensions, at three different levels, particularly in those areas where the main task/aim was based on connections and exchanges between the colonies (peripheral territories) and the empires (centres, 'metropolis', capitals). The first level referred to *travel*, as an instrument of transmission (perception) and/or knowledge (education, exchanges). The second level concerned the communication itself, as a means of transmission (publishing, diffusion) or as a medium (transport). Finally, the third level was the language understood as lexicon and as *possible world* (cultural totality).

Much has been said about 'travel writing' and, within this topic, about the so-called scientific authors. The two consecrated biographical models of Darwin and von Humboldt have generated a great number of approaches and theories regarding the ways in which the relationship between experiment (empirical science) and theory (theoretical science) has been developed and, equally importantly, how the scheme's centre/periphery and laboratory/publication was mutually defined and developed thanks to *travel*. Less a one-way imposition (from the empire to the colonies), this first level of translation showed a more complex scheme where the centre/empire was also affected at the level of its lexicon development and evolution.

From the middle of the nineteenth century, the development of communications began to be considered as an issue on a planetary scale (see Lebeau, 2005). Transport had already been in that situation since the Renaissance period, even though technically it was only possible to achieve full efficacy during the nineteenth century (see for example Mignolo, 2000 and 2003). What radically changed the perception of the past and the future was the fact that, increasingly, the movement of matter (goods, merchandising) was less attached to its financial dimension. Consequently, values increasingly acquired a non-physical dimension; they became abstract, and were transformed into *signs*. In this context, *translation* became more and more familiar on both a quantitative and a qualitative level. Transport was also radically affected by this change on the communicative level, as information was increasingly less attached to material objects (see Lebeau, 2005). Furthermore, in this situation interaction between scientific knowledge and technological evolution took place for the first time, a fact that constituted one of the most profound changes in Europe since medieval times (see Lebeau, 2005).

The increasing amount of travelling and the acceleration of exchanges generated a completely new domain: the belief that the future was something that had potentially already happened. Thus the emergence of this *horizon d'attente* as an element of daily perception opened the door to linguists and philosophers who started to consider language as the most appropriate tool for constructing/understanding these *possible worlds* and, equally importantly, as the only device for conceiving totalities – in which case the idea of 'science' was certainly affected. This condition gave support to another recent and more contemporary aspect: the regular need for a distinction between information ('información') and meaning ('significado'), or in other words, the critical acknowledgement that a human being makes a distinction and separation between what is immediately observable and what is not. In this case translation is constantly placed in a linguistic domain where the meaning of what can be said, of what is present, and of what is nameless, can be negotiated.

By the middle of the nineteenth century the issue of *translation* became an inevitable methodological aspect of any cultural or scientific activity; furthermore, academic authors began to develop theories around the idea of proper *conversion* tables and taxonomies (see for example Adas, 1989; and Clifford, 1997). Therefore, perspectives on historiography completely changed as authors were forced to become aware of the distinction between a *context of production* and a *context of reception* in relation to any particular value, concept or object (see for example Verón, 1987). In terms of the relationship between science and empire, this lead to increasing levels of detail and information in publications, but also to a new historical trend in the approach that soon became a new form of *historicism* (*Geschichtlichkeit*). This approach almost exclusively considered the relationship between science and empire from the chronological, historical ('historicity'), and time-scale point of view. At the same time, as a result, a *geo-epistemic approach* can be seen as an alternative and as an outside understanding of the question. By the end of the twentieth century it was almost widely accepted by specialists and academic authors that there were no longer any *original* contexts or *natural* situations within the relationship between science and empire, and that everything needed to be considered or thought about from the point of view of *translation*.

4. The *Translation Effect*

Given the above-mentioned conditions, *translation* can generally be understood, apart from its two dimensions (see point 3.2 and 3.3 above) and its three levels of engagement indicated earlier (see point 3.4 and Table 3.1 above), as translation in *language/linguistic* terms (T_1) and as translation in *cultural/scientific* terms (T_2). It is the close connection between these two forms that we can call the *translation effect*, that is, the fact that in the relationship between science and empire there is an exchange that goes directly to the conceptual/epistemological level. A schematised presentation of this situation can be found in Table 3.1 and Figures 3.1 to 3.3, as we will see.

The relevance of *translation*, bearing in mind the characteristics considered here, is that through its evolution it makes possible, as Latour (1987) and Montgomery (2000), among others, have made clear, the existence of a certain 'principle of relativity' or, better termed, a *principle of relationship*. In other words, any relationship, even if it is of political submission and domination, is an exchange that affects the conceptual basis of the relationship itself. And this is applicable not only to the relationship between empires and colonies but also to the relationship between science and empire. Therefore, for example, as we have already mentioned, the relationship between knowledge and science, and the bureaucratic and political *administration* of the relationship between science and the notion of empire, are part of the same problem.

Following Table 3.1, which compares all the elements mentioned about the idea of translation, we can see a combination of possibilities that can be considered as the space(s) within which the *principle of relationship* works. Comparing the two dimensions and the three levels of engagement, we can identify what type of translation is possible (T_1 and/or T_2). Within a material dimension and with travel as the level of engagement, the translation will be 'cultural' and/or 'scientific', just as it is when the dimension is material and the engagement grounded in communication. When the dimension is still material but the engagement is on the language level, then the translation is likely to be 'linguistic'. With an imaginary dimension, and in an engagement at the level of travel and communication, the translation will be 'linguistic' in both

	Travel (Level 1)	Communication (Level 2)	Language (Level 3)
Material dimension	T_1	T_2	T_1
Imaginary dimension	T_1	T_1	T_2

TABLE 3.1 The definitions of translation in the context of the science and empire relationship. T_1 stands for translation in 'linguistic' terms and T_2 in 'cultural/scientific' terms.

cases. However, in the case of engagement at the level of language, while the dimension is imaginary, the result is a 'cultural/scientific' translation.

More specifically, this *principle of relationship* which is made possible by the translation elements and characteristics can also be expressed in *spatial* terms. In Figure 3.1 we have attempted to show this. If we consider the three levels of engagement in terms of time (past, present and future) and the dimensions in terms of space (up and down, North and South, superior and inferior), then we can make a map of the relationship between empire and science expressed in a spatial form. The empires implement imperialism not only in relation to the colonies but they also manage to impose colonialism within the local elites (see Mignolo, 2005). Considering that language is the *ground zero* of the exchange between empires (centre/Capital) and the colonies (peripheral, periphery), and scientific language the most accurate expression of this relationship, then the implementation of this language as education (school) or as science (laboratory, applied science), is the classical way to exercise colonialism (see for example Dodson, 2005).

Consequently, from the point of view of the colonies, knowledge will be considered as a form of the past that has to inform education (schools), while laboratories and applied sciences will fit within the future and, in particular, within the notion of progress set up by the centrality of empire (see Figure 3.1). If the first relationship is characterized by the notion of historical transmission, the second is defined by the idea of communicating science and experiments. This whole situation can

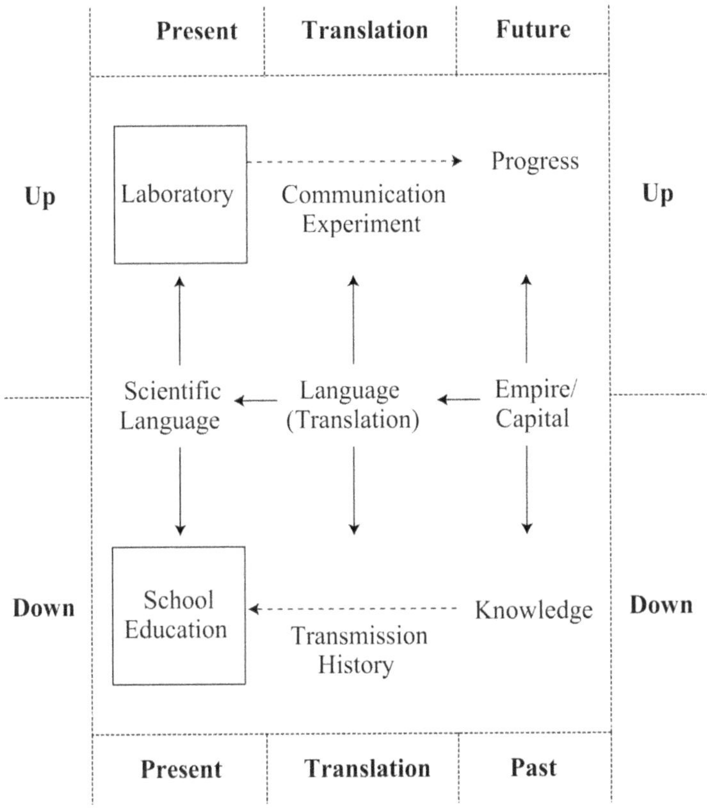

FIGURE 3.1. Translation 'in action'. Spatial displacement from the centre to the peripheral imperial lands.

also be read from the two dimensions of translation associated to a space category. Then, if we consider 'up' as imaginary and 'down' as material, it can be observed that in relation to knowledge and historical transmission, time works differently for the colonies and for the empire: knowledge considered within the empire is the past (tradition, culture, wisdom); however, when this knowledge is translated into the colonies it is considered as 'education' and as present (see Figure 3.1). Equally, the progress of the empire is understood as future, while translated into the colonies it is conceived as practical/applied science and as present. The knowledge transmitted to the colonies is 'imperial' since

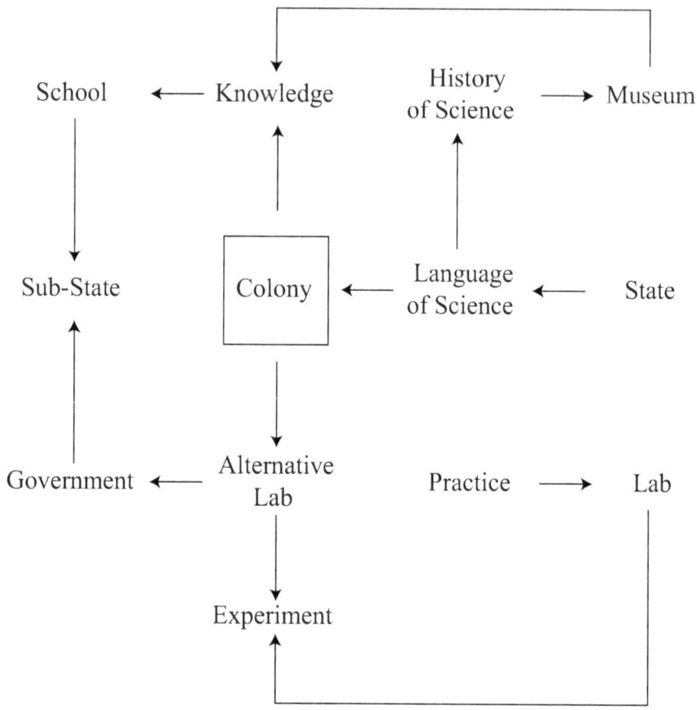

FIGURE 3.2. The main institutional exchanges between the empire and the colonies and vice versa.

it is conceived to work only in one direction. However, the role of the colonies is relevant as well, as this implemented applied science impacts upon the notion of knowledge itself and thus the colonies end up making a contribution to the conceptual evolution administrated by the empire's authorities and authors.

Figure 3.2 also expresses this situation, already described in Figure 3.1, but from an institutional point of view. In the same way that a direct connection between the imperial State and the colonial sub-State can be established (see Mignolo, 2005; and also Figure 3.2), there is a direct link between (i) the colonial development of schools, (ii) the location of museums and (iii) the establishment of the history of science as a discipline in the capitals and universities of the empire territory. Furthermore, the notion of the colonies as an *alternative* laboratory

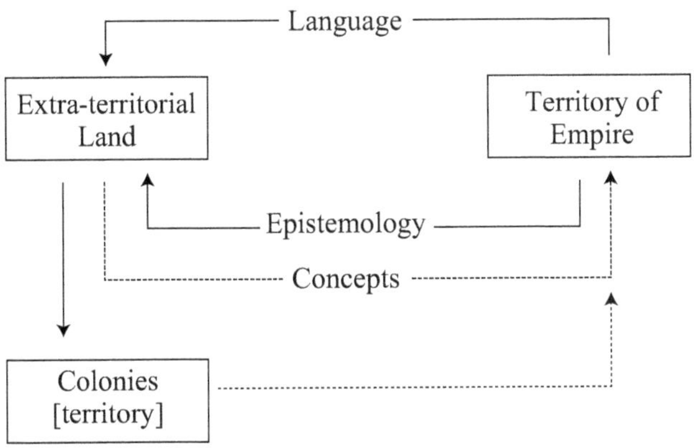

FIGURE 3.3 Diagram of the relationship between science and empire.

and/or *space for experimentation* was already present in the sixteenth and seventeenth centuries. And somehow the experimental expeditions of Malaspina, Darwin and Humboldt are the first step towards the more recent nuclear essays and pharmaceutical experiments conducted by corporations in ex-colonial territories.

Finally, Figure 3.3 illustrates an aspect that we consider to be the main point of this situation: there is an interaction between the Empire – Capitals, government, State, administration, academic institutions – and the colonies. This is founded on language, but functions primarily around the notion of knowledge, in which case what we call 'epistemology' can be understood as the area, or activity, that tries to understand the conceptual exchange between the central parts of the empires and their peripheral zones. The relationship between science and empire, once again, cannot be detached from the development of the colonies as such and, just as importantly, it cannot be detached from the constitution of a notion of science itself as 'imperial science'.

5. From Building the Empire to Constructing Science

Within the parameters of government, States and empires, science has always been considered as knowledge. Colonies do not have, from the predominant Western/European paradigm and perspective, any other period than that opened by the European 'modernity' in the fifteenth century (see Mignolo, 2003; also point 6 below). Moreover, in most cases, science has been considered as *the* paradigm of knowledge and it has been deployed as one of the principal tools to render the imperial domain vaster and more effective. In fact, the distinction between 'empire' and 'imperialism' can be made from the moment in which knowledge, as an equivalent of science, was used as an instrument of domination. While empire can be considered as having mainly an economic purpose, imperialism was also developing a political dimension. In this respect, it is impossible to approach the idea of 'modernity' within the Western/European world or culture without considering this relationship between empire and imperialism (see Mignolo, 2005). From the ideological point of view, modernity is the result of this implementation of knowledge on a State scale (see Schiebinger/Swan, 2004; also point 6 below).

The construction of empire(s) in Western/European domains cannot be separated from the development and exploitation of the colonies – people, land – while science and technology were among the most prominent instruments employed. The construction of houses, buildings, transport systems and engineering works within the colonies were the most powerful means that European science and technology used in order to become established (i) as an academic discipline, (ii) as a cultural device, and (iii) as a source of political legitimacy and governance. At the same time that, for example, railways and astronomical observatories were being built and designed in the colonies, the notion of science and technology, as discipline, as corpus, as *saber* and institution, was defining more and more *one sense* of reality and one set of methodological work considered as *natural* and definitive corpus. At the end of the nineteenth century, for example, Anglo-Saxon scientific journals dedicated more than half of their pages to 'colonial events' or to 'scientific events' connected with some colonial location (see for example Canaparo, 2003: 261–366).

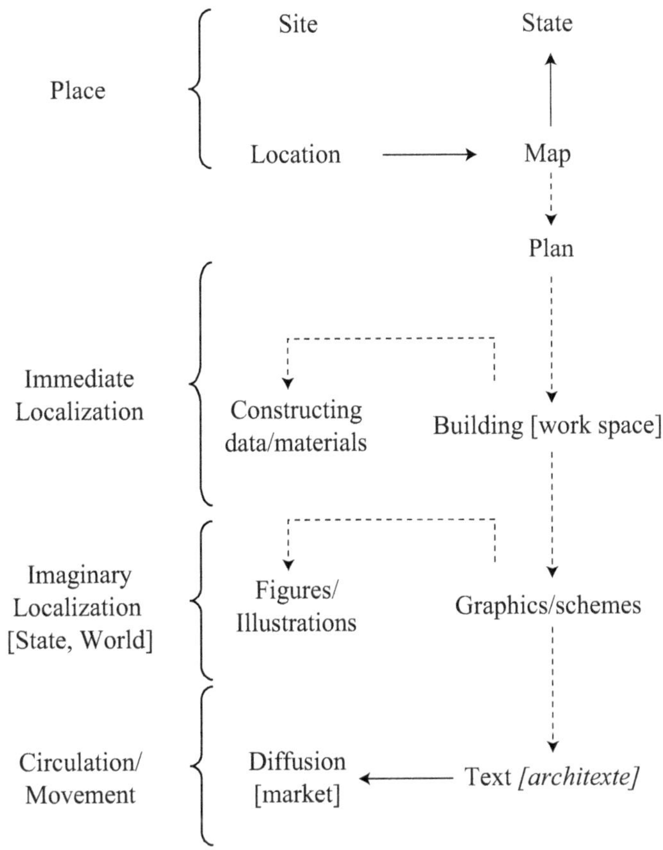

FIGURE 3.4. A possible formulation of the *imperial spatial matrix*.

Within this context, as we have already demonstrated, it is almost impossible to make a distinction between science and knowledge as two separate entities, and moreover, the nineteenth-century consideration of science as the most accurate and superior form of knowledge is certainly based on this imperial connection. This is why examining the *spatial* distribution and creation of an empire – rather than following chronological or temporal series or parameters – can help us to understand how *science as knowledge* has been determined and imposed. I will call this situation and adjustment the *imperial matrix of space*.

It is increasingly clear to specialists that the *perspective* employed in the construction of the Western/European empire(s) – in a physical and imaginary sense – is present in the culture(s) that developed within the colonies ('Creole societies') and is therefore present in the construction of science itself. This *continuum* is also what we call the *imperial matrix of space*. Interestingly, this situation clearly shows that, even when the general conditions regarding science and scientificity were established during the colonial period, the later development of these ex-colonial environments towards more modern organizations – countries, nations, etc. – confirms the fact that conceptual and technological evolution operates in both directions. The *imperial matrix of space* also refers to the influence and determinations/effects that the activities generated within the colonies have over the central/Capital domains of the empire. As Deleuze and Guattari point out, there is no such thing as an *inside and outside* of the empire's territory and/or land: what we find are *pliegues*, *insterstícios*, blind spots, *lignes de fuite* of the territory (Deleuze/Guattari, 1980). In fact, one astonishing finding after observing the works of authors and specialists in this area, is the regular insistence on always considering one *outside space* of the Empire, which actually becomes the *metaphysical exit* of the majority of the theories on the subject. Within the so-called 'colonial period' the empire is considered not only as symbolically allocated, but also as physically allocated. By contrast, within what we can call a 'post-colonial period' the empire – an imperialism to be more precise – is considered as non-visible, as 'signical' and *imaginary* (see Figure 3.4 and Figure 3.5). Given this context, the question of the empire's territory is less about the consideration of the colonies as separate entities because of their different political administration (independence, revolution, etc.), than about considering them as still being part of the empire's territorial space which had evolved from being symbolically and physically allocated ('colonial period') to a more non-visible and imaginary form and placement ('post-colonial period').

On the other hand, this mentioned *continuum* is questioned because there is a predominant *spatial dimension* present within the colonies, while, in the central/Capital parts of the empire, *chronologies and temporal approaches* ('temporalidad') are dominant: the logic of development of the colonies is different from the historiographical narrations established by the imperial scientific institutions, even when, as already stated, the imposition of a common *lexicon* is in place (see Mignolo, 2000; Harris, 1998; Nieto Olarte, 2000). And it is from this revisited discrepancy or, more precisely, from this discrepancy within a context of conceptual continuity, that the new theories on science and empire emerge (see for example MacLeod, 2000 and Mignolo, 2005).

Following then this *matrix*, it is possible to identify a sequence in the establishment of science as a *vehicular language* of the empire:

- Establishment of a Map = Space = Land = Territory.
- Regulation of the space in an imaginary sense = signs, symbols, codes.
- Regulation of the space in a terrestrial and physical sense = buildings, etc.
- Communication, movements and transport within the space = trips, travel and language.

The transformation of land into territories, through cartography, constitutes the first and most dominant device in establishing a sense of totality and standard authority within the first Western/European empires established by the Spanish. What must be acknowledged in this situation is not only the fact that this aforementioned association between knowledge and science generates a sort of Universal Reality, but also that it establishes a principle of reality for which 'real' becomes associated with the physical world (see for example Jardine et al., 1996; and MacKenzie, 1990). The empires, in the Western/European sense, were first established from an *imaginary* point of view: the introduction of an artificial *naturalization* of the land – sign, symbols, codes, taxonomies – was the first step towards the imposition of a *vehicular language* to interpret and read this 'nature' (see for example Arnold, 1996 and Nieto Olarte, 2000). In this respect, the imposition of science as a *vehicular language* in conceptual terms cannot be separated, as we have already stated, from the transformation of English into the common dominant idiom of governments, markets and academies.

Given these circumstances, it is predictable that the architectural organization of the State and of scientific organizations and institutions will follow a similar pattern within the origin of the colonies. Equally,

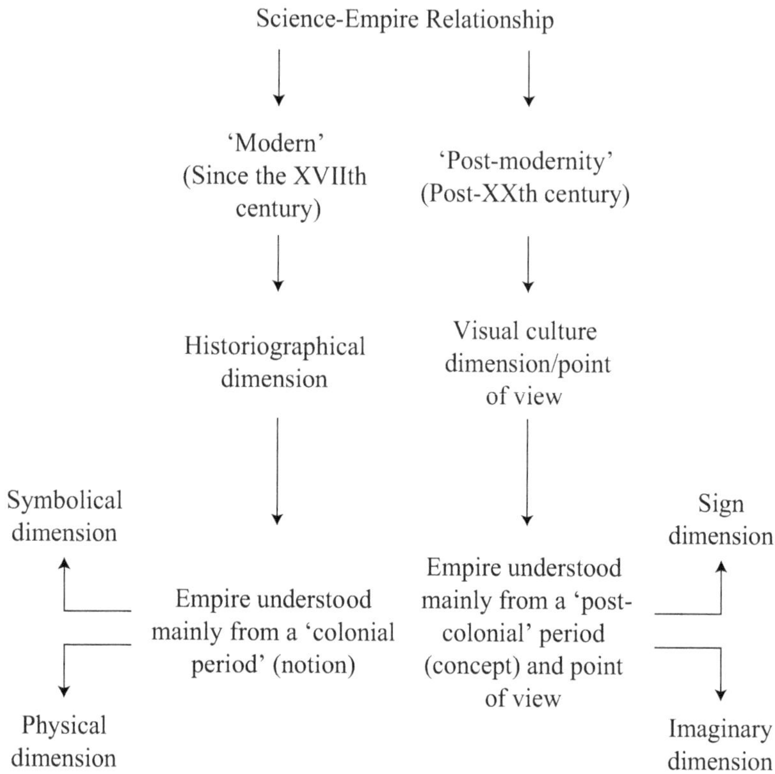

FIGURE 3.5. The relationship between science and empire from an evolutionary point of view.

the disposition of space employed to develop transport systems and communications, will follow an artificial sense of the space for which an extreme paradox will be created. If on one hand, the principle of reality is associated with the physical world, on the other hand, the 'physical world' is less of a pure material system, as one might at a first glance think, and more of an imposed artificial sense of nature and naturalization.

In Figure 3.4 we have set out an attempt to visualize this spatial/artificial situation – or scientific/naturalistic condition, as we might also call it – from which – as authors like Bruno Latour and others have studied already (see for example Latour, 1997; Latour/Woolgar, 1979)

– the initially mentioned *imperial spatial matrix* can be conceived more empirically. We do not understand this matrix as stable, fixed or unchanging or with the same invariable pattern. What is relevant in this original imperial scheme are the connections and *relations* that it creates. Rather than a fixed set of rules and characteristics, this matrix establishes a number of spatial relations that prevail over periodical and occasional changes (see point 3.3 above). This is where it is crucial to understand the association between knowledge and science, as this will allow us to understand why a number of authors postulate that (i) the development of modernity – first as an economic notion, then as a political perspective, and finally as a concept in ideological terms (see point 6 below) –, (ii) the growth of the State's bureaucracy and institutions, and (iii) the development of empires into *imperialism*, are all aspects of the same question and cannot be separated (see for example Mignolo, 2000 and 2005; Mandelssohn, 1976; Wallerstein, 1974). The localization of a particular territory has a direct connection with the 'localization of culture' (see for example Bhabha, 1994), a relationship that constitutes the ground for a notion of 'World' or 'imaginary', or, to use Edward Said's (1935–2003) expression, 'the World, the Text, and the Critic' (see Said, 1984). The relationship that exists between maps (cartography) and what we understand as territory (topography), is reproduced in the relationship between science and what we define as knowledge, just as it is reproduced in the relationship between the notion of reality and the physical space where scientific institutions are placed, and again in the relationship between scientific writings and the illustration and/or graphics made to legitimate/justify these writings for the market (Figure 3.4 and Figure 3.5). As Michel Foucault points out, what makes one idea or concept efficient and successful is not its immediate meaning but the relation between this particular meaning and the conceptual system (*epistémé*) to which it belongs (see Foucault, 1970). For that reason, the idea of the *imaginary* becomes increasingly relevant among authors as an expression which denotes a situation where, for example, a particular scientific building does not contain within its material existence the meaning under which is conceived everyday by the scientific practitioners that made use of it. The wonderful volume edited in 1999 by P. Galison and E. Thompson under the title of *The Architecture of Science* constitutes an interesting exploration in this direction.

As well as considering that the present *imperialism* no longer needs a notion of empire to function, we should consider the fact that *colonialism* does not need the conceptual certainty offered by the existence of traditional colonies in any specific location. Knowledge and subjectivity, as the main conceptual tasks for an actual geo-epistemic approach, can only function if we accept that the *scientificism* generated by the Western/European perspective no longer relies on a notion of 'Science' or 'Big Science' or 'Normal Science' but rather on a *conceptual language* that needs to be revisited. The recent attention paid by many authors to travel, movement, imaginary constructions and visual fabrications, among other elements, proves in our view the pertinence of this *conceptual language* (see for example Riffenburgh, 1994; Willis, 1998; and Clifford, 1997).

6. Modernity, Science, Knowledge

From the historiographical point of view, three different conceptions of 'modernity' can be considered in the relationship between science and empire. When modernity is understood as beginning in the fifteenth century/Renaissance period, it is mainly considered from an *economic* point of view, that is, initially linked to the origin of the idea and implementation of *capital* in Europe. Secondly, modernity is also considered as originating in the eighteenth century, founded on a *political* point of view, when *contractual* ideas about societies and the related political theories appeared in Europe, which coincides with the transformation of capital into *capitalism*. Thirdly, from the *ideological* point of view, modernity is considered as originating in the nineteenth century and, equally importantly, as an autonomous *concept* and not just as an idea.

In this way, the evolution of the relationship between science and empire, as shown in Figures 3.5 and 3.6, can be understood as follows: the idea of modern/modernity is entangled in the Western/European sensibility with the notions of empire and notions of empire and science, such that the three periods, economical, political (institutional) and ideological, can be applied to the notion/concept of empire (Figure

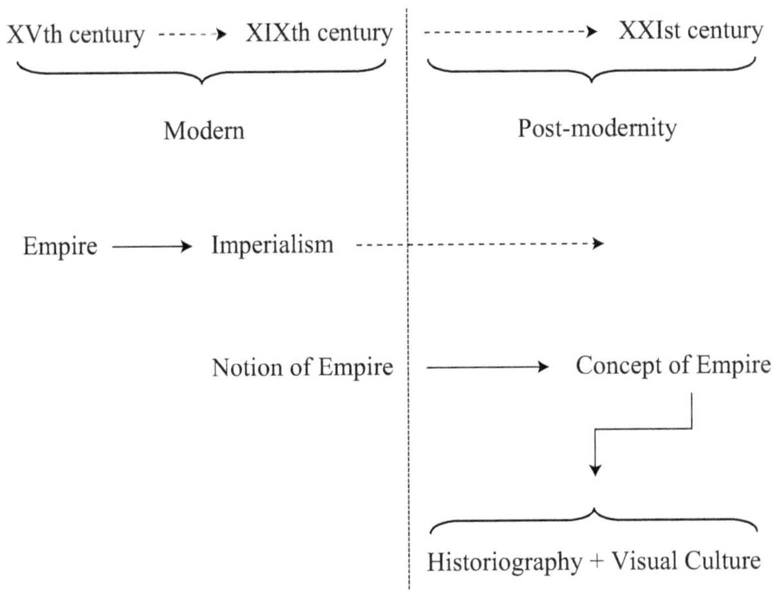

FIGURE 3.6. The historical evolution of the notion/concept of empire.

3.6) and to the idea of science (Figure 3.7). If the concept of empire(s) is placed at the centre of colonialism itself then, similarly, the European idea of science ('scientificity') is situated at the centre of knowledge as such. The central role of the *imaginary* in reproducing a number of colonial aspects within the colonies (see Mignolo, 2005) cannot be detached from the crucial function of science/scientificity in reproducing a number of European characteristics of knowledge within the territory and institutions of the colonies.

The complexity of the argument can be observed thus: the relationship between science and empire not only refers to a reciprocal relation between central parts of the empire ('relaciones fuertes') and its borders, but also to a regular connection between the peripheral territories on their own ('relaciones débiles') – relations that for the colonies can have crucial relevance, as authors like Eduardo Ortiz (1935–) have pertinently argued (see Ortiz, 1995 and 1998). The direct connection between modernity, colonialism and knowledge also explains why, from the nineteenth century, when the so-called post-

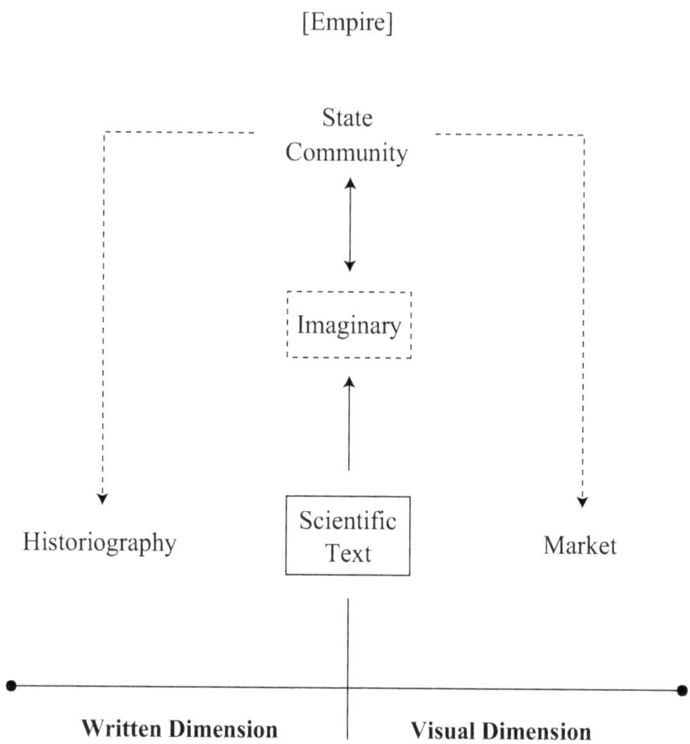

FIGURE 3.7. The two dimensions of the concept of empire.

colonial period began in Western/European societies, i.e., when the majority of the colonies started to gain political independence from the Spanish Empire, a second and *more subtle form* of colonialism emerged. It was not by chance that this new colonialism appeared at the same time that modernity was beginning to be considered as *ideology*. This 'second' colonial perspective was developed among the natives and *créole* inhabitants of the colonies (see Ortiz, 1995 and 1998). Paradoxically, the imperial scientificism that dominated colonial space (see for example Babini, 1954) was never more *foreign* than when the *local* governments began to assert their political independence from the European powers. The case of Latin America seems astonishingly clear in this respect (see for example Ortiz, 2005). As shown in Figures 3.5

and 3.6, a connection can be established with the previously mentioned situation (see Table 3.2) for which the so-called post-colonial period is not understood as the disappearance of colonialism, but on the contrary, as representing the emergence of a second *imaginary* form of it.

A further consideration can also be made in this respect. Within the Western/European tradition, imperialism is entangled with the changes in economics and the system of transport which transformed *comercio* (commerce) and capital into capitalism and financial capital from the seventeenth century. This is why imperialism is normally associated with the second and third forms/types of modernity mentioned above (see also Figure 3.8). Max Weber's (1864–1920) analysis of the modern State is exemplary in this sense (see for example Weber, 1968). In this way, colonialism can be considered as a previous stage of imperialism and this can be closely observed in such cases as Spain which, being an empire during the sixteenth century, did not survive as such into the eighteenth and nineteenth centuries and never managed to transform its colonialism into imperialism (see for example Pimentel, 2000; Elena/Ordoñez, 2000). The Spanish authorities' attachment to a *written* dimension when describing/approaching the colonies is another confirmation of this condition (see Figure 3.7).

One aspect which needs further consideration is the fact, following the previous comment, that a frequent misunderstanding is generated regarding the meaning of 'post-coloniality' (see Table 3.2). That is to say, a distinction is not always made between the *notion* of 'post-coloniality', when it is used to describe an historical event, and the same notion when used as *concept*, or, in other words, to indicate a particular approach to the question of Empires, colonialism and imperialism (see Figure 3.5). If we consider empire from the 'central' point of view, there is a clear evolution from a colonial form of domination towards independent control of the colonies and ex-colonies (the historical notion of post-coloniality). But if we consider empire from a marginal/peripheral perspective, what we have is the transformation of colonialism into a *mentality*, as explained above, that will not become embroiled with imperialism until later in the nineteenth and twentieth centuries. It is relevant to consider here that the political independence of the colonies from the European powers did not always mean an evolution away from colonialism, which means that 'post-colonialism' is not a useful term to describe what happened within the colonies from the nineteenth century onwards. The already mentioned case of Latin America is illustratative of this situation. Colonialism was still in place even during the formation of the 'National States' and imperialism started to operate much later with the shift of the imperial 'centre' from Europe to the United States at the beginning of the twentieth century, and, no less significantly, with the shift from a physical and State-orientated form of domination to another more symbolic, abstract, financial and corporative one.

In these circumstances a paradoxical situation occurs in certain domains such as, to keep the same example, Latin America. Scientific activities were more relevant in relation to central parts of the empire during the colonial period than later, after the introduction of imperialism. The *colonial* situation of the scientific activities during the period when the 'National States' were founded was based on two aspects: first on *a material culture* (practices, experiments, etc.) and, secondly, on an *enlightened* notion of science. By contrast, with the introduction of imperialism, science and scientific activities become an *abstract and symbolic* exchange of information and objects. And, once again, there is a paradox here: from the local point of view the central parts of the empire were more accessible and *familiar* within the colonial period than afterwards, once imperialism had been introduced. This is why, within certain areas such as South and North America, during the colonial period authors were setting up/introducing the question of science in the context of 'civilization vs. barbarism', as well as in the context of *study* and intellectual formation. At the beginning of the twenty-first century this paradoxical situation once again becomes a visible issue as the post-industrial societies of the periphery fall further and further outside the scientific market. This market was designed and controlled for the scientific and publishing corporations as a new form of monopolistic domain not only for scientific activities and their development, but also for the notion of knowledge and understanding in general.

7. Towards a Scientific Imaginary and an Invisible Empire: From Historiography to Visual Culture

One very interesting thing that emerges while looking at the evolution of the notion of empire and its links with science is that, when the argument appeared during the period of the Great European expeditions, cartography was one of the most relevant tools for providing a perspective and/or panorama on the meaning of empire and, equally importantly, on how science can help to develop the perspective of empire. Cartography was the dominant tool in the construction of geographical spaces: constructing maps and charts was equivalent to constructing a geographical and topographical space. From the point of view of meaning, the map then was the territory.

Today we face a similar situation: the notion of empire has been replaced by markets or global markets or, more precisely, the classical administrative and State-orientated notion of empire has been replaced by more efficient and more obvious forms of imperial dominance. However, the requirement for a visual dimension, in cartographic terms, is still there (Figure 3.4 and Figure 3.5). The 'return' of cartography as a way of constructing a *universo de discurso* seems to be a matter on which several authors agree (see for example Gregory, 1994 and Godlewska/Smith, 1994). Moreover, a *cartographic thinking* seems to be required in order to be able to speculate on all the transformations

Colonial period	**Post-colonial period**
Modernity as economic or political notion.	Modernity as ideology
Imperialism as policy, as material reality and as government.	Imperialism as 'mentality' and as *imaginary*
Notion of Empire	Concept of Empire
Symbolic/physical dimension	'Sign'/Imaginary dimension

TABLE 3.2. The transformation of colonialism from a conceptual point of view.

through which 'imperial science' has been transferred from State-orientated institutions and academies (like the Royal Society of London) to private corporations (like the Institute for Scientific Information of the Thompson Corporation in Washington or the journal *Nature* of the German Holtzbrinck Group in London) (See for example Canaparo, 2003: Chapter 11).

The relation between science and empire nowadays, outside of the academic consideration developed within Anglo-Saxon universities, highlights the need to confront two major connected issues. In a culture where the principle of reality and the justification of the existence of matter rely on non-visible elements, the relevance of the concept of *imaginary* has become part of the notion of science itself (Figures 3.6 and 3.7). Equally, as we have already stated, the conceptual and empirical evolution of the notion and dimension of empire becomes something radically different and its common and distinctive character is its *invisibility*: the elements that inform the empire(s) – or, more precisely, the 'imperial dimension' – are not visible in the traditional empirical way. At the most they are *signs* in a semiotic sense (see for example Eco, 1978, 1980 and 1984; and also Figure 3.5).

The *historiographical operation* that authors like W. Mignolo rightly indicate as typical from the empirical-historical construction of colonies – that is, imposing *naturalization* in relation to an *artificially* constructed past – has currently evolved into a similar attempt to constantly (i) *re-write the past* and (ii) *establish the limits of the present in relation to the past*. By contrast with previous attempts, however, it does so within the context of a dominant visual culture (see for example Mitchell, 1980 and 1994). Writing is no longer the *zero degree* of culture: *images* now function in the way that writing did before. In this case, as Paul Virilio and other European thinkers have pointed out, the consumption and consummation of the past and the present can happen very quickly, and so the once incipient sense of *horizon d'attente* is now pushed to the extreme: expectations about the present become more and more important and more relevant than other factors (see for example Virilio, 1984; and Jeudy, 1990). Consequently, it is easier to understand the *turn* generated within the study of science and scientific knowledge, produced by the so-called 'Science Studies' approach, as it introduced a methodology that considers science to be placed at the heart of a visual

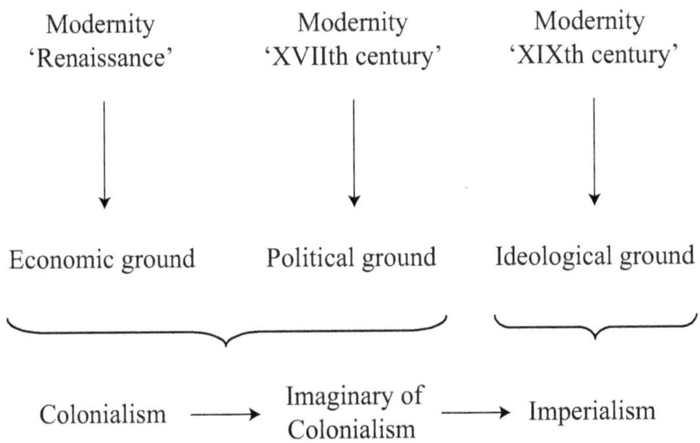

FIGURE 3.8. The evolution of colonialism into imperialism. The case of Latin America.

culture. Contrary to what might appear, the recent debates concerning the status of science or the status of scientific knowledge are less connected with a particular notion of science than with the relation to the consideration of an actual and crucial epistemological point. The idea of 'reality' has been replaced by the notion of *imaginary* and the dimension of 'matter' has been replaced by a space built on *images*.

The above-mentioned pertinence of cartography in the present time, or, more precisely, the approach that we can describe as *cartographic thinking*, is based on two things. It is based on the one hand, on the already mentioned discrepancy between the temporal perspective deployed in the central part of the empire and the spatial perspective used in the *fronteras* of the empire. On the other hand, it is based on the disappearance of the traditional sense of *land* or *homeland*: 'topographic amnesia', as Paul Virilio called it (see Virilio, 1988). This is not only a consequence of the 'post colonial' and post-industrial development of central European countries, it also corresponds to a consequence of a certain logical development of coloniality. The principles whereby in the colonies, and in the nations and/or countries that developed subsequently, land was transformed into *territories* using 'scientific elements' have ceased to be operative. At present the possibility of referring knowledge or subjectivities to those territories understood in a traditional way is questioned: objects, individuals, corporations and academic institutions correspond to an extreme mobility (perception, knowledge) and to a new sense of emigration/immigration (movement, locality, situation). The notion of geo-epistemology that is proposed here tries to acknowledge and address this issue.

8. Bibliography

Adas, M. *Machines as the Measures of Men: Science, Technology and Ideologies of Western Dominance* (Ithaca: Cornell University Press, 1989).

Alam, A. 'Imperialism and Science', in *Social Scientist*, vol. 6, No.5, December, 1977.

Andersson, J. S. *How to define 'Performative'* (Stockholm: Libertryck, 1975).

Arnold, D. (ed.). *Imperial medicine and indigenous societies* (Manchester: Manchester University Press; New York: St. Martin's Press, 1988).

Arnold, R. *The Problem of Nature: Environment, Culture and European Expansion* (Oxford: Oxford University Press, 1996).

Austin, J.L. *How to Do Things with Words* (Oxford: Clarendon Press, 1962).

Baber, Z. *The science of empire: scientific knowledge, civilization, and colonial rule in India* (Albany: State University of New York Press, 1996).

Babini, J. *La evolución del pensamiento científico en la Argentina* (Buenos Aires: Raigal, 1954).

Barton, T. *Power and Knowledge: Astrology, Physiognomics, and Medicine under the Roman Empire* (Ann Arbor: University of Michigan Press, 1994).

Basalla, G. 'The Spread of Western Science', in *Science*, vol. 156, 1967, pp. 611–622.

Bayly, C. A. *Imperial Meridian: The British Empire and the World* (London: Longman, 1989).

Bell, M., Butlin, R. A., and Heffernan, M. (eds.). *Geography and Imperialism, 1820–1940* (Manchester: Manchester University Press, 1995).

Bhabha, H. K. (ed.). *Nation and Narration* (London: Routledge, 1990).

———. *The location of culture* (London: Routledge, 1994).

Bonneuil, Ch. 'Development as Experiment: Science and State Building in Late Colonial and Postcolonial Africa, 1930–1970', in *Osiris*, vol. 15, 2000, pp. 258–281.

Breman, J. (ed.). *Imperial Monkey Business: Racial Supremacy in Social Darwinist Theory and Colonial Practice* (Amsterdam: VU University Press, 1990).

Burckhardt, J. *The Civilization of the Renaissance* (London: George Allen and Co., 1914).

Canaparo, C. *Ciencia y escritura* (Buenos Aires: Zibaldone, 2003).

Cañizarez-Esguerra, J. *How to Write the History of the New World: Histories, Epistemologies, and Identities in the Eighteenth-Century Atlantic World* (Stanford: Stanford University Press, 2001).

———. 'Iberian Colonial Science', in *Isis*, vol. 96, 2005, pp. 64–70.

Cardoso, F. E./Faletto, E. *Dependency and Development in Latin America* (Berkeley: University of California Press, 1979).

Chambers, D. W./Gillespie, R. 'Locality in the History of Science: Colonial Science, technoscience, and Indigenous Knowledge', in *Osiris*, vol. 15, 2000, pp. 221–240.

Cipolla, C. M. *European Culture and Overseas Expansion* (London: Penguin, 1970).

Clifford, J. *Routes: Travel and Translation in the late Twentieth Century* (Cambridge, Mass.: Harvard University Press, 1997).

Darwin, Ch. *Voyage of The Beagle* (London: Penguin, 1989).

Deans-Smith, S. 'Nature and scientific Knowledge in Spanish Empire Introduction', in *Colonial Latin America Review*, vol. 15, num. 1, June 2006.

Deleuze, G./Guattari, F. *Mille Plateaux* (Paris: Minuit, 1980).

Deleuze, G./Guattari, F. *Qu'est-ce que la Philosophy*? (Paris: Minuit, 1991).

Dodson, M. S. 'Translating science, translating Empire: the power of language in colonial North India', in *Comparative studies in society and history*, vol. 47, num. 4, 2005, 809–835.

Drayton, R. *Nature's Government: Science, British Imperialism and the Improvement of the World* (New Haven: Yale University Press, 2000).

Eco, Umberto. *A Theory of Semiotics* (Bloomington: Indiana University Press, 1978).

———. *Signo* (Barcelona: Labor, 1980).

———. *The Sign of Three: Dupin, Holmes, Peirce* (Bloomington: Indiana University Press, 1984).

Edney, M. H. *Mapping and Empire: The Geographical Construction of British India, 1765–1843* (Chicago: University of Chicago Press, 1997).

Elena, A. and Ordoñez, J. 'Science, Technology, and the Spanish Colonial Experience in the Nineteenth Century', in *Osiris*, vol. 15, 2000, pp. 70–82.

Elsner, J. and Rubies, J. (eds.). *Voyages and Visions: Towards a Cultural History of Travel* (London: Reaction, 1998).

Feingold, M. (ed.). *Jesuit Science and the Republic of Letters* (Cambridge, Mass.: MIT Press, 2003).

Figueiroa, S./da Silva, C. 'Enlightened Mineralogists: Mining Knowledge in Colonial Brazil, 1750–1825', in *Osiris*, vol. 15, 2000, pp. 174–189.

Fischer, K. P. *History and Profecy: Oswald Spengler and the Decline of the West* (New York: Peter Lang, 1989).

Foucault, M. *The Order of Things* (London: Tavistock Publications, 1970).

Fulford, T./Kitson, P. J. (eds.). *Travels, Explorations and Empires: Writings from the Era of Imperial Expansion 1770–1835* (London: Pickering and Chatto, 2001–2002).

Galeano, E. *Open Veins of Latin America* (London: Monthly Review Press, 1973).

Galera Gómez, A. *La ilustración española y el conocimiento del nuevo mundo: las ciencias naturales en la expedición Malaspina, 1789–1794* (Madrid: CSIC, 1988).

Galison, P./Thompson, E. (eds.). *The Architecture of Science* (Cambridge, Mass.: MIT Press, 1999).

Gascoigne, J. 'The Royal Society and the emergence of science as an instrument of state policy', in *British Journal for the History of Science*, vol. 32, 1999, pp. 171–184.

———. *Science in the Service of Empire (1998). Joseph Banks, the British state and the uses of science in the age of revolution* (Cambridge: Cambridge University Press, 1998).

Godlewska, A./Smith, N. (eds.). *Geography and Empire* (Oxford: Blackwell, 1994).

Gorra, M. *After Empire. Scott, Naipaul, Rushdie* (Chicago: Chicago University Press, 1997).

Greene, J. C. *Darwin and the modern world view* (Baton Rouge: Louisiana State University Press, 1961).

Gregory, D. *Geographical imaginations* (Oxford: Oxford University Press, 1994).

Hall, Catherine. *Cultures of Empire: A Reader – Colonisers in Britain and the Empire of the Nineteenth and Twentieth Centuries* (Manchester: Manchester University Press, 2000).

Harding, S. *Is Science Multicultural? Postcolonialisms, Feminisms, and Epistemologies* (Bloomington and Indianapolis: Indiana University Press, 1998).

Hardt, M./Negri, T. *Empire* (Cambridge, Mass.: Harvard University Press, 2000).

Harris, S. J. 'Long-Distance Corporations, Big Sciences, and the Geography of Knowledge', in *Configurations*, vol. 6, 1998, pp. 269–304.

———. 'Mapping Jesuit Science: The Role of Travel in the Geography of Knowledge', in O'Malley, J. et al. (eds.), *The Jesuits: Cultures, Sciences, and the Arts, 1540–1773* (Toronto: Toronto University Press, 1999), pp. 216–222.

———. 'Jesuit Scientific Activity in the Overseas Missions, 1540–1773', in *Isis*, vol. 96, 2005, pp. 71–79.

Harrison, M. 'Science and the British Empire', in *Isis*, vol. 96, 2005, pp. 56–63.

———. *The Tools of Empire: Technology and European Imperialism in the 19th Century* (New York: Oxford University Press, 1981).

Headrick, D. R. *The Tentacles of Progress; Technology Transfer in the Age of Imperialism* (New York: Oxford University Press, 1988).

Hefferman, M. J. 'A State Scholarship: The Political geography of French International Science during the Nineteenth Century', in *Transactions of the Institute of British Geographers*, vol. 19, 1994, pp. 21–45.

Hobsbawm, E. J. *The Age of Empire, 1875–1914* (New York: Pantheon Books, 1987).

Humboldt, A. *Personal Narrative* (London: Penguin, 1995).

Hunt, B. J. *Doing Science in a Global Empire: Cable Telegraphy and Electrical Physics in Victorian Britain* (Chicago: University of Chicago Press, 1997).

Huntington, S. P. *The Clash of Civilizations: And the Remaking of World Order* (New York: Free Press, 2002).

Jardine N., Secord, J. A./Spary E. C. *Cultures of Natural History* (Cambridge: Cambridge University Press, 1996).

Jassanoff, S. 'Biotechnology and Empire: The Global Power of Seeds and Science', in *Osiris*, vol. 21, 2006, pp. 273–292.

Jeudy, J-P. *Le désir de catastrophe* (Paris: Aubier, 1990).

Kant, I. *Universal Natural History and Theory of the Heavens* (Edinburgh: Scottish Academic Press, 1981).

Kumar, D. *Science and Empire: Essays in Indian Context, 1700–1947* (Delhi: Anamika Prakashan, 1991).

Lafuente, A. 'Enlightment in an Imperial Context: Local Science in the Late-Eighteenth-Century Hispanic World', in *Osiris*, vol. 15, 2000, pp. 17–30.

Latour, B. *Science in Action* (Cambridge: Harvard University Press, 1987).

———. *Petit leçons de sociologie des sciences* (Paris: Seuil, 1997).

Latour, B./Weibel, P (eds.). *Making Things Public. Atmosphere of Democracy* (Cambridge, Mass.: ZKM/MIT Press, 2005).

Latour, B./Woolgar, S. *Laboratory Life* (Milton Keynes: Open University, 1979).

Lebeau, A. *L'engrenage de la technique - Essai sur une menace planétaire* (Paris: Gallimard, 2005).

Livingstone, D. N. *Putting Science in its Place. Geographies of Scientific Knowledge* (Chicago: Chicago University Press, 2003).

Lopes, M. M. and Podgorny, I. 'The Shaping of Latin American Museums of Natural History, 1850–1990', in *Osiris*, vol. 15, 2000, pp. 108–118.

Lyotard, J-F. *The Differend: Phrases in Dispute* (Minneapolis; University of Minnesota Press, 1988).

Mackay, D. *In the Wake of Cook: Exploration, Science, and Empire, 1780–1801* (London: Croom Helm, 1985).

MacKenzie J. M. (ed.). *Imperialism and the natural world* (Manchester: Manchester University Press, 1990).

———. *The Empire of Nature: Hunting, Conservation and British Imperialism* (Manchester: Manchester University Press, 1998).

MacLeod, R. 'Passages in Imperial Science: From Empire to Commonwealth', in *Journal of World History*, vol. 4, 1993, pp. 117–150.

———. 'On Visiting "Moving Metropolis": Reflections on the Architecture of Imperial Science', in Reingold, N. and Rothenberg, M. (eds.), *Scientific Colonialism: A Cross Cultural Comparison* (Washington: Smithsonian Institution Press, 1987).

———. 'Nature and Empire: Science and the Colonial Enterprise – Introduction', in *Osiris*, vol. 15, 2000, pp. 1–13.

Mandelssohn, K. *Science and Western Domination* (London: Thames and Hudson, 1976).

Marsden, B. *Engineering an Empire: Technology, Science and Culture, 1760–1914* (Basingstoke: Palgrave Macmillan, 2003).

McClellan, J. E./Regourd, F. 'The Colonial Machine: French Science and Colonization in the Ancien Regime', in *Osiris*, vol. 15, 2000, pp. 31–50.

Merson, J. 'Bio-prospecting or Bio-Piracy: Intellectual property Rights and Biodiversity in a Colonial and Postcolonial Context', in *Osiris*, vol. 15, 2000, pp. 282–296.
Mignolo, W. D. *Local Histories/Global Designs* (Princeton: Princeton University Press, 2000).
———. *The Darker Side of the Renaissance: Literacy, Territoriality, and Colonization* (Ann Arbor: University of Michigan Press, 2003).
———. *The Idea of Latin America* (London: Blackwell, 2005).
Mignolo, W./Schiwy, F. 'Double Translation', in Maranaho, T./Strek, B., eds. *Translation and Ethnography. The Anthropological Challenge or Intercultural Understanding* (Tucson: University of Arizona Press, 2003), pp. 3–29.
Miller, D. Ph./Reill, P. H. (eds.). *Visions of Empire: Voyages, Botany, and Representations of Nature* (Cambridge: Cambridge University Press in association with the UCLA Center for Seventeenth- and Eighteenth-Century Studies and the William Andrews Clark Memorial Library, 1996).
Minguet, Ch. *Alexandre de Humboldt: historien et géographe de l'Amérique* (Paris: Maspero, 1969).
Mitchell, W. J. T. *The Language of Images* (Chicago: Chicago University Press, 1980).
———. *Picture Theory: Essays on Verbal and Visual Representation* (Chicago: Chicago University Press, 1994).
Montgomery, S. L. *Science in Translation* (Chicago: Chicago University Press, 2000).
Needham, J. *Science and Civilization in China* (Cambridge: Cambridge University Press, 1954–1986). 6 volumes.
Nickles, Thomas. 'Philosophy of Science and History of Science', in *Osiris*, 10 (1995): 138–163.
Nieto Olarte, M. *Remedios para el imperio: Historia natural y la apropiación del Nuevo Mundo* (Bogotá: Instituto Colombiano de Antropología e Historia, 2000).
Ortiz, E. 'A Convergence of Interests: Einstein Visit to Argentina in 1925', in *Ibero-Amerikanisches Archiv*, volumen 21, números 1–2, 1995, pp. 67–126.
———. 'Army and Science in Argentina: 1850–1950)', in Forman, P/Sánchez Ron, J. M. (eds.), *National Military Establishments and the Advancement of Science and Technology* (Dordrecht: Kluwer, 1996), pp. 153–184.
———. 'The Transmission of Science from Europe to Argentina and its Impact on Literature: from Lugones to Borges', in Fishburn, E. (ed.), *Borges and Europe Revisited* (London: ILAS, 1998), pp. 108–123.
———. 'Científicos y Ciencia en la Argentina de 1860: El affaire Santiago Cáceres', in *Boletín de la BCN*, Buenos Aires, número 122, 2005, pp. 41–64.
Ortiz, E/Rubinstein, H. 'Impacto de los países escandinavos en el desarrollo de la física en la Argentina: 1960–2000', in *Ciencia Hoy*, Buenos Aires, volumen 15, número 18, 2005, pp. 38–45.
Osborne, M. A. 'Acclimatizing the World: A History of the Paradigmatic Colonial Science', in *Osiris*, vol. 15, 2000, pp. 135–151.
———. 'Science and the French Empire', in *Isis*, vol. 96, 2005, pp. 80–87.

Palladino, P./Worboys, M. 'Science and Imperialism', in *Isis*, vol. 84, 1993, pp. 91–102.
Parker, A./ Kosofsky Sedgwick, E. (eds.). *Performativity and Performance* (London: Routledge, 1995).
Petitjean, P./Jami, C./Moulin, A-M. (eds.). *Science and Empires: Historical Studies about Scientific Development and European Expansion* (Dordrecht: Kluwer Academic, 1992).
Pimentel, J. *La física de la monarquía: Ciencia y política en el pensamiento colonial de Alejandro Malaspina (1754–1810)* (Madrid: Doce Calles, 1998).
———. 'The Iberian Vision: Science and Empire in the Framework of a Universal Monarchy, 1500–1800', in *Osiris*, vol. 15, 2000, pp. 17–30.
Polanco, X. *Naissance et development de la science-monde: Production et reproduction des communautés scientifiques en Europe et en Amérique Latine* (Paris: La Decouverte, 1990).
Pratt, M. L. *Imperial Eyes: Travel Writing and Transculturation* (London: Routledge, 1992).
Pruna, P. M. 'National Science in a Colonial Context: The Royal Academy of Sciences in Havana, 1861–1898', in *Isis*, vol. 85, 1994, pp. 412–426.
Pyenson, L. *Cultural Imperialism and Exact Sciences: German Expansion Overseas 1900–1930* (New York: Peter Lang, 1985).
———. 'Pure Learning and Political Economy: Science and European Expansion in the Age of Imperialism', in Visser, R. P. W. et al. (eds.), *New Trends in the History of Science* (Amsterdam; Rodopi, 1989).
———. 'Cultural Imperialism and Exact Sciences Revisited'. In *Isis*, vol. 84, 1993, pp. 10–108.
Raj, K. 'Colonial Encounters and the Forging of New Knowledge and National Identities: Great Britain and India, 1760–1850', in *Osiris*, vol. 15, 2000, pp. 119–134.
Reingold, N./Rothenberg, M. (eds.). *Scientific Colonialism: A Cross Cultural Comparison* (Washington: Prentice Hall and IBD, 1987).
Riffenburgh, B. *The Myth of the Explorer: The Press Sensationalism and Geographical Discovery* (Oxford: Oxford University Press, 1994).
Roszak, Th. *Where the Waste Land Ends* (Garden City: Anchor Books, 1972).
Ryan, J. R. *Picturing Empire: Photography and the Visualization of the British Empire* (London: Reaktion Books, 1997).
Said, E. W. *The World, the Text and the Critic* (London: Faber and Faber, 1984).
———. *Culture and Imperialism* (New York: Knopff, 1993).
———. *Orientalism* (London: Penguin, 1995).
Schiebinger, L. 'Forum Introduction: The European Colonial Science Complex', in *Isis*, vol. 96, 2005, pp. 52–55.
Schiebinger, L./Swan, C. (eds.) *Colonial Botany: Science, Comerse and Politics in the Early Modern World* (Philadelphia: University of Pennsylvania Press, 2004).
Searle, J. R. 'How performatives work', in *Linguistics and Philosophy* 12, 1989, pp. 535–558.

Semonin, P. 'Empire and extinction: The dinosaur as a metaphor for dominance in prehistoric nature', in: *Leonardo: International Journal of the Contemporary Artist*, 30 (1997): 171–182.

Sheets-Pyenson, S. *Cathedrals of Science: The Development of Colonial Natural History Museums During the Late Nineteenth Century* (Kingston: McGill-Queen's University Press, 1988).

Shils, E. 'Centre and Periphery', in Polanyi, M. (ed.). *The Logic of Personal Knowledge* (London: Routledge and Keegan Paul, 1961).

Simpson, Ch. (ed.). *Universities and Empire: Money and Politics in the Social Sciences during the Cold War* (New York: The New Press, 1998).

Smith, M. R./Marx, L. (eds.). *Does Technology Drive History? The Dilemma of Technological Determinism* (Cambridge, Mass.: MIT Press, 1994).

Snow, C. P. *The Two Cultures and the Scientific Revolution* (Cambridge: Cambridge University Press, 1959).

———. *The Two Cultures; and, A Second Look : An Expanded Version of 'The Two Cultures and the Scientific Revolution'* (Cambridge: Cambridge University Press, 1964).

Sorlin, S. 'Ordering the World for Europe: Science as Intelligence and Information as Seen from the Northern Periphery', in *Osiris*, vol. 15, 2000, pp. 51–69.

Spengler, O. *The Decline of the West. Form and Actuality* (London: George Allen and Unwin, 1926).

———. *The Decline of the West. Perspectives and World History* (London: George Allen and Unwin, 1928).

Steiner, G. *In Bluebeard's Castle: Some Notes towards the Re-definition of Culture* (London: Faber and Faber, 1971).

Verón, E. *La semiosis social* (Barcelona: Gedisa, 1987).

Virilio, P. *L'horizon négatif* (Paris: Galilée, 1984).

————. *La machine de vision* (Paris: Galilée, 1988).

Wallerstein, I. *The Modern World System: Capitalist Agriculture and the Origins of the European World Economy in the Sixteenth Century* (New York: Academic Press, 1974).

Weber, M. *Economy and Society* (London: Bedminster Press, 1968).

Willis, C. (ed.). *Building the Empire State* (New York: Norton, in association with the Skyscraper Museum, 1998).

Worboys, M. 'The Colonial World as Mission and Mandate: Leprosy and Empire, 1900–1940', in *Osiris*, vol. 15, 2000, pp. 207–218.

Young, R. M. *Darwin's Metaphor: Nature's Place in Victorian Culture* (Cambridge; Cambridge University Press, 1985).

Zeller, S. 'The Colonial World as Geological Metaphor: Strata(gems) of Empire in Victorian Canada', in *Osiris*, vol. 15, 2000, pp. 85–107.

Ziólkowski, M. S./Sadowski, R. M. (eds.). *Time and Calendars in the Inca Empire* (Oxford : B.A.R, 1989).

Žižek, S./Daly, G. *Conversations with Žižek* (London: Polity, 2004).

Chapter 4
Language, Knowledge and *horizon d'attente*

> What occurs in a living system is analogous to what occurs in an instrumental flight where the pilot does not have access to the outside world and must function only as a controller of the values shown in his flight instruments. His task is to secure a path of variations in the readings of his instruments, either according to a prescribed plan, or to one that becomes specified by these readings. When the pilot steps out of the plane he is bewildered by the congratulations of his friends on account of the perfect flight and landing that he performed in absolute dark. He is perplexed because to his knowledge all that he did at any moment was to maintain the readings of his instruments within certain specified limits, a task which is in no way represented by the description that his friends (observers) make of his conduct. (Maturana/Varela, 1980: 51)

Nowadays academic authors face two major problems. The first is the fact that all cultural elements, including the intellectual analysis of culture itself, constitute *values* that, in a greater or lesser proportion, can be accounted for in terms of a financial scheme. And this situation is as valid for aesthetic manifestations as it is for philosophical speculations. The second problem or question is the fact that all cultural elements, including the definition of culture itself, have a *historiographical past* and an *encyclopaedia of meaning* that cannot be avoided. While the first problem is associated with concepts such as the market, media, communications, etc., the second concerns notions like society, understanding, science, etc. These two major problems are faced in similar way, more or less directly, by a great deal of academic authors regardless of their discipline or domain.

The issue here is not to point out this already known distinction but to indicate its consequences. A great many intellectual and speculative analyses of Latin America fail to acknowledge the fact that whatever revolutionary perspective you might have regarding past or present situations, all perspectives are forced to deal with these two major obstacles. One of the most astonishing cases to my knowledge perhaps is that of Walter Mignolo's excellent book *The Darker Side of the Renaissance* (1995). The sometimes brilliant analysis that Mignolo offers of the meaning and dimension of writing in the pre-colonial and colonial period of Latin America, for example, fails incredibly to achieve any practical or functional perspective on the present expressed in the form of the two above-mentioned obstacles. From the contemporary point of view, the very illuminating diagnosis of concepts and conditions of knowledge in the pre-colonial and colonial periods of Latin America do not offer any practical perspectives about the future and viability of this analysis. It is true that the author does not, at any point, express any intention or interest in doing so. However, given the two major problems indicated before, and in whatever form they could be considered, this perspective and orientation towards the future ('porvenir') seems inevitable for the reader, as some reviews of this work have already indicated (see for example Vera Tudela, 1997 and Barnes, 1998). Mignolo's analysis implicitly emphasizes the results of a relationship between what Reinhart Koselleck (1923–2006) called 'the space of experience and the *horizon d'attente*' (Kosellek, 1986: 300 ff.).[1] Mignolo's solid and coherent analysis exposes the deficiencies of any theory of knowledge that ignores the *local transformations* (see for example Mignolo, 2000: Chapter 1), but curiously does not explore this in relation to his own proposed analysis.

And this is where we see the consequences of this double problem: all conceptual or cultural analysis of Latin America *ipso facto* faces a question of accountability and functionality. And it is not a matter of agreement or disagreement, nor is it an ideological aspect to agree or

1 The production of history as a *present phenomenon* (even when it concerns the past) but developed in *fictional terms* from the epistemic point of view is an issue that Mignolo raised many times: the modern sense of *historicity* is part of Western/European colonialism (see for example Mignolo, 2000: 127 ff.). Furthermore, Mignolo, expressing his ideas about 'peripheral thinking' reveals the present extreme conditions of Kosellek's hypothesis: '… l'époque moderne est caractérisée non seulement par un rétrécissement de l'espace d'expérience, qui fait que le pasée parait toujours plus lontain à mesure qu'il parait plus révolu, mais par un écart croissant entre l'espace d'expérience et l'horizon d'attente.' (Ricoeur, 1985: 384). This is a perspective that, according to Mignolo himself is present in other European philosophers, like J. Derrida and G. Deleuze (see for example Mignolo, 2000: 72 ff.)

disagree with, it is a matter of *media* existence (communications, mass media) and market circulation (finance, values). And not merely in simple commercial terms of exchange but also in the way that academia and universities (as places of 'knowledge') operate with regard to the subject. The *locus of enunciation* mentioned earlier (see Chapters 1 and 2) – which also acts here as 'culture' (first problem) – is already established and well-defined, and, as is the case with all subjects that are normally determined conceptually by 'historiography' (second problem), it also constitutes their understanding and legitimacy within this context of media existence and market circulation.

However, less important than establishing what is already obvious to the reader, i.e., that the definition of language and knowledge is relevant to determining the meaning of a particular place, is the indication of the fact that language and knowledge have an interactive, mutual self-definition in relation to what we would like to call, in the words of P. Ricoeur (1985: 375–391) and R. Koselleck (1986: 300–322), *horizon d'attente*.[2] Furthermore, from this perspective, the heuristic of culture – or a principle of understanding the culture in Latin America – should be related less to the origins of European culture itself than to a biological theory about knowledge (see for example Maturana, 1995b).[3] The understanding of culture nowadays is more *epistemic* than ever in the sense that the privileges that historical, sociological or political interpretations enjoyed in the past have now disappeared

2 According to Koselleck, 'L'orizzonte si referisce a quella linea dietro la quale si chiude, in futuro, un nuovo spazio di esperienza, che peraltro non è ancora visibile'. (Koselleck, 1986: 306). And Ricoeur comments: 'Trois thèmes se détachent des soigneuses analyses sémantiques de Koselleck. D'abord, la croyance que l'époque présente ouvre sur le futur la perspective d'une *nouveauté* sans precedent, puis la croyance que le changement vers le mieux s'accélère, enfin la croyance que les hommes sont de plus en plus capables de *faire* leur histoire. Temps nouveau, accélération du progrès, disponibilité de l'histoire – ces trois thèmes ont contribué au déploiement d'un nouvel horizon d'attente qui, par choc en retour, a transformé l'espace d'expérience dans lequel se sont déposés les acquis du passé.' (Ricouer, 1985: 379).

3 This is where, as indicated in Chapter 2, some authors avoid the discussion of culture or, even the concept of culture itself, because it is impossible to grasp any practical or functional meanings from it.

(see for example Knorr-Cetina, 1999).[4] And this is where the real transformation against traditional European colonialism is happening: the *reivindicaciones* (claim, demand) of the pre-Columbian past are less connected to a particular content or subject than to the fact that knowledge is more and more understood in biological and *languaging* terms,[5] to use H. Maturana's well-known expression (see Maturana/Varela, 1980; Maturana, 1995a and 1995b).[6] Despite the need to deal with the two obstacles already mentioned, cultural evolution should be considered as an issue where biology and *languaging* get together (see Maturana, 1995b). Traditional institutional or historical explanations cannot confront the condition of the present in Latin America, even when they may help us to understand a particular local aspect:

> El conocimiento es un fenómeno biológico y sólo puede ser entendido como tal; cualquier discernimiento epistemológico en el dominio del conocimeinto requiere de este entendimiento. (Maturana, 1995b: 208)

The fact that the cultural evolution (understood as general epistemic development of concepts and understanding) of what we call 'Latin America' is, in our view, less related nowadays to a particular ideology or a specific idea of language than to a general perception based on biology and finance, is something that indeed radically questions an important number of interpretations made of/about Latin America

4 'I prefer to think that if comparative processes need an origin, it – like many others – should not be looked for among the Greeks but among some of the features common to living organisms. Discerning differences by identities, and vice versa, also seems to be a feature of human intelligence, and consequently, so are comparative attitudes. Why should comparatism be interpreted as a Greek invention instead of as a human need for (better) adaptation and survival?' (Mignolo, 1995: 19).

5 '...y hacemos lo que hacemos como seres humanos en el lenguaje y lenguajeando (*languaging*)' (Maturana, 1995a: 39).

6 It is in this context that we can understand the lack of any paradigm or, more precisely, the complaint ('queja') of many local authors about the lack of general perspective or paradigm. The impossibility of ascertaining a general view about Latin America, contrary to mainstream understanding (see for example King, 2004), is less related to a multiplicity of conflicts or realities than to an absence of concepts that can refer such totality.

or any related subject in the present.[7] In this chapter we will briefly explore a view of Latin America, understood as thinking perspective (see Chapters 1 and 2), via this above-mentioned approach.

1. The Idea of Knowledge

In Chapter 1 we considered knowledge in terms of 'thinking' to establish a distinction between 'local' and Western/European ('not local') forms of knowledge. We also presented the connection between 'space' and 'thinking' as a way of exploring the hypothesis that in pursuing the 'location of knowledge' there is something different to the traditional historical or scientific pattern. Associations with notions of translation, material culture and *imaginatio* were also made. In Chapter 2, following on from the previous ground, we focused on the ideas of environment, cinematography, language and writing, and their relationship to knowledge and/or thinking. Furthermore, we again explored the notion of 'space' and 'thinking', particularly in relation to the terms/notions of land and territory. Finally, in Chapter 3, the notion of knowledge was explored in relation to science, empire and a dominant 'Enlightenment' paradigm of understanding. The ideas of politics, geography, modernity, reality and colonialism, among others, were explored. To conclude this *tour* we will now try to briefly focus on two or three further aspects to finalize our considerations of the notion of knowledge.

Knowledge, like other concepts connected with culture and/or society, has been considered in/for Latin America as a term connected with philosophy or, ultimately, with the history of philosophy. Contrary to this, and from the perspective we are/will be proposing here, knowledge has two dimensions: one is *biological* and the other is about *logistics*.

7 Perspectives about environment, food and present material practices can in this sense become much more useful than historical or ideological perspectives at the moment of explain, for example, why the colonial project established in the nineteenth century is not what it was intended to be and why the countries/Nations created at that time have become an artificial entities more than ever.

Knowledge, understood as thinking is defined by its epistemological conditions, by its boundaries. And this is where space and thinking have a natural connection: space defined by its boundaries and peripheral areas matches thinking conceived from its limits and ultimate conditions (see for example Mignolo, 2000: 91–126). Furthermore, this natural connection is rooted in a biological condition as understood by Humberto Maturana: if any study of the transcultural conditions that characterize the local is possible it should be based on 'biology of cognition' terms (see Maturana/Varela, 1980: 5 ff.). This is for at least two reasons: (i) because through it 'reality' is understood as a result of a community consensus construction, where 'community' is understood not only in the sense of Ferdinand Tönnies (1855–1936) (1955) but also in the sense of Imre Lakatos (1922–1974) (1978), and (ii) because any notion of 'nature' or objectivism is considered as constructed (see for example Glasersfeld, 1997). Furthermore, the biology of cognition understands knowledge (i) as a *deriva*, (ii) as something constructed locally and for a specific purpose, (iii) as *uniquely* related to an observer, (iv) as an *autopoietic* entity but subject to constant interactions, and (v) as expressed in terms of *languaging* (see Maturana/Varela: 1980). And all these generic aspects, which basically constitute the foundation of 'biology of cognition', also constitute an ideal general approach to the understanding of the local culture.

1.1 The Escape to Invisibility

One element that deserves further consideration is the fact that authors or specialists – like the previously mentioned W. Mignolo – considered only one definition of colonization, even when referring to many forms (see for example Mignolo, 2000: 127 ff., and 172 ff.). The point rightly addressed by Mignolo concerning the fact that European philosophers missed certain aspects of colonialism in their theories (see for example 2000: Chapter 1), is somehow paradoxically duplicated when he considers *only one definition* of colonialism, i.e. that which (i) refers to countries/societies or (ii) is anthropologically rooted in a dimension of the human.

A further consideration of the definition of colonialism is required here. As we have discussed elsewhere (see Canaparo, 2005b: 294 ff.; and also Canaparo, 2007: 238–247) and also stated in this book (see Chapters 2 and 3), one of the most radical changes within the colonies (of Western/European dominance), that helped to destroy any perspective or paradigm of 'realism', was the transformation of matter into an element by definition *invisible*. Furthermore, the disappearance of a relationship – in terms of understanding – between visibility and invisibility is considered as a starting point of a new revolutionary period (see for example Virilio, 1988 and 1989; also Baudrillard, 1981: *passim*).

And it is this *way to invisibility*, which embraces concepts and realities, that makes the sense of colonialism even more radical than that envisaged by Mignolo and others. We are already living in an era where colonialism affect bodies, artefacts and objects. The colonialism of people, nations, countries and ethnicities has certainly not disappeared but has been pushed further away by this new colonial environment that cannot be understood using the classical points of reference. And technology – in the broad sense understood for example by André Lebeau (2005) – is the main instrument/tool of this 'new colonialism'. The frantic speed of this development and the general financial scheme attached to it will radically change (i) our sense of bodies – starting with the classical seventeenth-century anatomic sense of the human body – ('bodying', 'embodiment', *corporalización*), (ii) the functionality of artefacts (*artefactalidad*) and (iii) the definition of objects (*objectualidad*).

And this is where I struggle to deal with the 'aftermath' of fascinating theories such as those originating from W. Mignolo. Despite their criticisms of the Western/European notion of knowledge – 'knowledge' itself, to be more precise – and of the scientific/enlightenment paradigm, I honestly believe they do not produce a working alternative and, in some respects, remain caught up in what they criticized. In addition, the issue of 'loci of enunciation' itself, for which the notion of knowledge conceived by European academic authors in the sixteenth century – or authors with European academic training, like the majority of the Jesuits – is considered as relevant, while the current equivalent within US academy is ignored. Why should the notion of knowledge that Spanish institutions had in the sixteenth century be more relevant than the fact that Latin America – as a concept and cultural entity – is nowadays a dominant subject produced not locally but within US academia?

1.2 The Domain of Logistics

In peripheral areas thinking is not about rationality but rather about *logistics*. What is relevant is not a particular concept or scheme but the way in which this concept or scheme became what it is or, no less relevantly, the way in which we became familiar with them. Everything in this sense is *operational* (see for example Plata, 2007) in peripheral thinking (see for example Mignolo, 2000: 91–126). The condition of knowledge – as we can call it in classical terms – within peripheral areas does not refer to the validity of a particular argument but to the dominating practical elements of such a process of validation – the art of managing imaginary worlds, languages, objects, goods and displacements is the truly speculative root of peripheral thinking. Furthermore, if peripheral thinking needs to be defined – further to what has already been written by authors like Mignolo or Dussel (see for example Mignolo, 2000: 91–126 and Dussel, 1998: *passim*) – it can be done using this 'managing' element as the driving element. It is not a question of being able to explain the historical evolution of the concept of Latin America. What is relevant here is not so much being able to deploy Latin America as a concept but rather the ability *to view* Latin American as a concept within a context dominated by finance values and logistics parameters.[8]

And somehow, as the European P. Virilio stated (see Virilio, 1984 and 1996), it is the late development of capitalism and finance that also places logistics or, more precisely, the 'logistics of perception' at the

8 Some authors, following the theories of Immanuel Wallerstein (1930–) propose the *economization* of colonialism as a way of re-formulating the interpretation of the impact of two different cultures clashing (see for example R. Grosfoguel et al., 2002).

centre of what is possible from an intellectual point of view.[9] Logistics is linked to movement (politics: *dromology*), to cinematography (sociology, philosophy: *imaquinación*) and to the ever-increasing (*in crescendo*) invisible core of matter (science, scientificism: 'epistemic cultures') because all these domains are ones that allow us to manage our practical, conceptual and immediate sense of environment.[10]

It is paradoxical in this context that an author like P. Virilio, for example, with a radical and critical perspective about naturalism and/or any traditional view of empiricism, should have much in common with Maturana, whose perspective is based on 'neurofisiología de la percepción'. The 'perception' understood by Virilio as visual, cybernetic, and technological orientated has then a 'natural' ground as understood by Maturana, that is, as a constructed element. The perception is not a function, is not a quality, and is not an attribute acquired *ab initio*. On the contrary, is a task performed by the body and as such is constantly constructed in the interaction between the body and the environment.[11]

The financial and epistemic condition of today's public life and ordinary activities in Western/European domain directly influences the way in which the local culture itself has been explained and considered for most of the twentieth century. The impossibility of the majority of authors and analysts separating this formal and epistemic condition from the conception of culture itself certainly has important consequences for their ability to conceive new ways of thinking or, to be more precise, for their ability to consider the possibilities of other non-European spaces. M. Foucault was quite right in identifying the conceptual power behind

9 By 'logistic of perception' we understand the general assumption (i) that our perception of the environment is mainly visual, (ii) that it is mediated by technological devices, and (iii) that perception is crucial in determining the sense of understanding (see Virilio, 1984).

10 In *El imaginario Patagonia* (forthcoming) we discussed how the domains of movement, cinematography and conceptual evolution (epistemology) constitute the core of any contemporary explanation in the analysis of peripheral areas ('zonas coloniales').

11 According to Virilio, 'The logistics of perception began by encompassing immediate perception (…). The logistics of perception was from the start the geographic logistics of domination from an elevated site.' (Virilio, 1997: 3).

the economic ideas of the seventeenth century (see Foucault, 1966), as was P. Bourdieu, years later, in identifying their consequences in terms of values and institutions (see for example Bourdieu, 1992). In both cases, however, as indicated earlier in relation to European authors, a non-Western/European space surfaced only as an *ontological* subject (see Bourdieu, 1988).

1.3 Writing and the Environment

In this context, written knowledge as a concept is considered (i) as a Western/European product (see for example Derrida, 1967; also Mignolo, 1994b) or (ii) as a development of an historical context – for example in relation to modernity (see for example Benjamin, 1983; Benjamin, 1991; Frisby, 1986; and Sarlo, 1992) or to the development of the modern State (see Weber, 1925; also González Casanova, 1984). However, the peripheral condition of Latin America still remains clear. Therefore the historical or philosophical European roots of the local thinking are less relevant than the present environmental determination and condition. In this sense thinking became a logistic dimension: the transcultural nature of Latin America (Spitta, 1995) is directly connected with the interdisciplinary condition of logistics – and here it is worth pointing out that 'discipline' does not refer to academic subjects – considered by many authors as 'colonial' (see for example Mignolo, 1995: 29–68) – but to *market areas*: the effectiveness of academia and university subjects (in the Western/European sense), in the construction/discussion of the local environment and principles of identity, is almost non-existent. The huge dimension of the university population and the universal influence within the local cultural and political systems is due to the elements that make it dysfunctional in relation to the European model. And this happens in such a paradoxical way that, contrary to what was envisaged by the 'founding fathers' of the nineteenth century, newly created nations, local universities, instead of being reproduced according to European model of knowledge, education and, ultimately, of society, became dysfunctional towards any non-local paradigm. The impossibility of establishing a *habitus* within academia in Latin America (in the European sense) goes back at least to the beginning of the twentieth century (for an explanation of the European notion of *habitus* see Bourdieu, 1984). The imitative component regarding European trends or perspectives is completed distorted by the *voyage*: theories, ideas and interpretations travel quite frequently today and their context of production is detached from their context of reception, to put it in classical terms (see for example the thesis of F. Cusset regarding the 'French Theory': Cusset, 2003).

1.4 Localization and Understanding

The question of 'location of knowledge' already mentioned in this work is neither new nor recent. As an *argument* it can be tracked down to medieval times (see for example Zumthor, 1993). As an *academic subject*, however, it emerged more recently. The fact that knowledge could be related to a notion of culture, to a notion of language or to geographical locations, also has a long connection in historiography with the evolution and definition of knowledge (see for example Cassirer, 2004). However, what we believe is even more recent in relation to knowledge is the fact that 'science', 'reality' and 'universalism' are no longer considered *ipso facto* as the natural partners of knowledge. And this situation has (i) an ontological aspect – that continues to be discussed – and (ii) a epistemological dimension – that is not discussed any more because it is assumed to be already realized.[12]

In these circumstances W. Mignolo is convincing when he proposes a *correction* for the main philosophical European propositions made in the twentieth century, even when they were consistently radical concerning more formal traditions (see Mignolo, 2000: 49–90). Mignolo mainly discusses the cases of G. Deleuze (Mignolo, 2000: 78–84 and 201–202) and J. Derrida (Mignolo, 2000: 64 ff.), and the proposed correction consists in the fact that these authors – although having

12 The ontological question is about the *nature* of our World in relation to human beings, and the epistemic question refers to the status of scientific knowledge or, more precisely, to the consequences that the 'triumph' of so-called scientific knowledge might have on the idea of knowledge in general.

radical views on many European issues – are missing the point about *colonialism* (see Mignolo, 2000: 197 ff.). The works of Paul Virilio are not in this sense an astonishing exception within the Western/European context, yet Virilio's notion of vision, space and architecture push him closer to the consideration of the actual meaning of 'colonial space' (see Virilio, 1995), even when he did not specifically discuss 'colonialism' as such – something that Derrida, for example, did explicitly (see Derrida, 1996).

2. The Idea of Language

Michel Foucault's well-known work, *Les mots et les choses* (1966), presents a very straightforward explanation of the shift that, according to the author, happened in Western/European culture around the sixteenth century in Europe and from which a modern notion of language was introduced. The basic point is that since then language has become a *tool with the ability to see and name in an abstract way* and, therefore, with the capability to represent objects without having the material object itself. This signical, semiotical and symbolical feature of language, according to Foucault, did not exist before in Europe.[13]

The increasing development of language as an abstract tool, following the Western/European specialization of language understood as alphabetic writing (see for example Derrida, 1967) was, according to Foucault, *en germen* in the changes that occurred within the sixteenth century. More relevant for our purpose is to notice that this shift towards the abstract has developed together with an increasing definition of the language elements as signs – the development of semiotics, for example, as an academic discipline from the time of Charles Sanders Pierce onwards certainly responds to this situation. It also explains the dominant role of English within Western/European culture as a

13 As has been discussed elsewhere (Canaparo, 2005a and 2007), these capabilities will have another *turn* when language starts to deal with what is not visible but has meaning in everyday life. We will return to this point later in this chapter.

language that can best express the taxonomy and lexicon of finance and technology, the two driving elements of Western/European civilization since the seventeenth century.

However what I find most fascinating about this work is not necessarily its diagnosis of European knowledge or the conceptual evolution that it proposes, but the way it implicitly addresses the future as something that will be less connected to any kind of notion of reality or history and more related to an imaginary and argumentative *construction* of the present. The explanation of how the notion of *representation* features at the core of the notion of language within Western Europe during the sixteenth century has a strong and somehow unexpected actuality that not many authors envisaged in 1966 when Foucault's work was first published. And, less than focusing on questioning/criticizing the notion of representation itself – something that has already been done by many authors (see for example Canaparo, 2003: 859 ff.) – what is significant is the *turn* that Foucault's description of the main features of language has, in the sense that the current situation in peripheral areas like Latin America is one where the representative capabilities of language have undergone yet another change on top of the one indicated by Foucault for the European domain. This *double turn* is what is sometimes so difficult to explain to non-European analysts.

According to Foucault, during the sixteenth century words and the objects (or things) started to be separated from one another according to their semantic dimension. Words and things started to be clearly split, even if related, according to their meaning. They started to be classified and their relations developed a taxonomical dimension which in two centuries, according to Foucault, generate (i) the conception of the naturalistic sense of the World, (ii) the mechanical sense of physical organization, and (iii) the evolutionary theory of life and art (see Foucault, 1966: chapter 2). The lack of any distinction between words and objects that was characteristic from the period previous to the Renaissance, according to Foucault, was then transformed by the introduction of an *abstract condition of language*. And if we introduce this argument here it is because we believe peripheral areas like Latin America have started to generate a knowledge that supposes a sort of *second turn of language* by which the abstract condition already discussed by Foucault, which initially only referred to language in the form of signs, also started to refer to physical objects, to the material world itself. This radicalization of the connections by which language operated in European terms resulted in the fact that not only the link between language and words is abstract and artificial but also that the physical dimension of objects and things themselves has become an abstraction and an artificial entity.[14] The relationship and exchanges between visibility and invisibility which, according to Foucault, are based on what he calls 'similitudes', opened the way to a new notion of language in the sixteenth century and gave it the sense of reality and semantic efficacy. Nowadays we do not have that past functionality and efficacy and a *radical turn to invisibility* has taken place. As ever the reality and legitimacy of language has been founded on a completely invisible set of concepts and ideas of matter. And in these circumstances the collapse of the major part the local political thinking is inevitable, as many of the recent crises of legitimacy in different Latin American governments have proven more than once.

Language today in the peripheral space of Latin America is not a linguistic entity in the sense described by Foucault ('La prose du monde') but a culture issue in the way discussed here: language is a varied set and collection of signs and objects whose 'materiality' does not have immediate meaning. What is more, language is a varied set and collection of signs and objects that are radically detached from words and from any fixed form or system of meaning. More specifically, language is a set of images whose possible narration does not have any standard or predictable form. This change in the epistemic valuation of alphabetic writing in relation to the location of culture is what has enabled many of the recent post-colonial interpretations of Latin American. And, from this perspective, Western/European tradition is not relevant, for the locals, because of its intellectual qualities or arguments, but because of its strategic value and its historiograpical situation. And this also explains why, when considering any Greek or Latin traditions, for example, the reference is less linked to any cultural characteristics or traditions (see for example Canaparo, 2005b: 165–196), as some authors

14 This constitutes the base for what we will later call the *travel towards invisibility*. We will return to this point.

might think, but more to an inevitable historiographic dimension in the sense mentioned before in this chapter (see also Chapter 2).

The aforementioned switch in the conception of language – towards a visual domain but also towards an *invisible* dimension – could also be related to the fact that a strict alphabetic and written form of understanding of language – as was the case in the past – is no longer useful for dealing with an idea of culture or knowledge. And this is when the notion of *languaging* ('lenguajeando'), as *in principle* described by H. Maturana (1995a and 1995b), fits with the language condition of what we have been considering as located ('localizado') in Latin America in terms of knowledge. Maturana characterizes language as an operational tool that makes all human activities possible. Furthermore, language always functions in a transitive state, connecting observers, concepts, interpretations and actions (see Maturana, 1995a: 19 ff.). Language here functions in an evolutionary way although not considering evolution as a principle or starting point but as the *result* of a process (Maturana, 1995a: 152). The human being, understood as a knowledgeable entity, cannot be considered outside language, in which case language works entirely as an *epistemological domain* that coordinates the physical and non-physical worlds. Therefore, language is also the instrument capable of generating (i) consensus, (ii) cooperation in a communicative sense and (iii) coordination of actions. *Languaging* is a 'forma de vida' (way of life) based on regular interactions that operate as coordination and consensus (Maturana, 1995a: 41). *Languaging* is what prevents the material systems from being absolutely mute:

> Lo central al fenómeno humano es que se da en el lenguaje. Y lo central del lenguaje es que sólo en él se dan la reflexión y la autoconciencia. El lenguaje es un sentido antropológico es, por lo tanto, el origen de lo humano propiamente tal a la vez que su caída y liberación. El lenguaje saca la biología humana del ámbito de la pura estructura material, e incluye en ella el ámbito de la estructura conceptual al hacer possible un mundo de descripciones en el que el ser humano debe conservar su organización y adaptación. (Maturana, 1995a: 16)

And this is where the distinction between academic authors and local critics occurs: while authors like W. Mignolo, for example, see the concept of 'colonialism' as explaining and representing one 'estado de cosas' (state of the question, state of the art) that we can call Latin

America, perspectives like Maturana's regard concepts as constructions that do not represent anything since they are the 'estado de cosas' itself. And it is for that reason that Maturana maintains that 'lo genético no determina lo humano, sólo funda lo humanizable' (Maturana, 1995a: 15). Latin America is, in this context, a contemporary phenomenon, whose characteristics only acquire the status of knowledge and culture when they can be formulated as a coordination of actions and as a coordination of a particular consensus.

And, in this context, it is possible to maintain that the important thing is not only to rescue the aboriginal languages and giving them real social status, not only freeing the Castilian language from its remaining Peninsular academic sickness. Above all it is relevant to determine the legitimacy of the Castilian-Latin American as an intellectually capable language – which, contrary to what some authors might think (see for example Fuentes, 2004), is far from actually what happens. Writing in Castilian within the academic and publishing market is a political and strategic decision that does not prevent authors and intellectuals from learning and expressing themselves publicly in English (or any other vehicular language), as we are doing here. What is at stake here, as Mignolo (2005) has correctly pointed out already, is the fight to impose a 'paradigm of conviviality' and this will not happen if we expect it to come from the colonial and imperial use of languages that Anglo-Saxon universities and publishers have introduced in the development of the global culture that determines our sense of contemporaneity nowadays.

3. The Thinking, The Factual, and the World

Following the meaning of thinking discussed before (see Chapter 1) we can now make conjectures about how the sense of environment is built up. If thinking is a way to revert to the philosophical assumptions of Western/European tradition, what was traditionally considered as 'factual' is now, within this peripheral situation, functioning as an ontological perspective and therefore as something that has no epistemological weight in the local environment and its development.

On the contrary, the notion of 'world' that historiographically speaking is connected with the origin of what we understand as Latin America (see for example Mignolo, 2005: x–xx), has been expanded and, as a core concept, now has many formulations that can hardly fit with the univocal meaning of universalism proposed by Western paradigm of enlightenment. In this sense, the world is no longer a fixed framework but a construction and *ad hoc* development. To be more precise, the notion of development in connection with Latin America is always an *ad hoc* phenomenon.[15]

The direct connection that linked a universal sense of reality and knowledge with a *unique* conception of reality and the world, is what can no longer be sustained. Latin America as an entity, contrary to what some authors still believe (see for example Galeano, 1986, 1987a and 1987b; and also Fuentes, 1969), is not founded on these Western/European parameters of objectivism and worldness. It is founded rather on more circumstantial and changing aspects that refer on the one hand to a regular and recurrent set of material and on the other hand to imaginary elements.

Our discussion about speculation, geography and the construction of territories as developed in Chapter 1, fits with this contemporary absence of a universal framework of understanding. Equally, the notions of local, loci of enunciation, reversal thinking, naturalism, and translation acquire their local dimensions precisely because of this *absence*.

In Chapter 2, the ideas of space, geometry and ground were developed even further in relation to this lack of universal perspective, both in an imaginary and a physical sense. The notion of 'environment' in its local dimension was also a way to confirm the difficulties or 'real' possibilities of such a universal understanding. Furthermore, on

15 That is also why we do not share many of the viewpoints that attempt to make Latin America a 'special cultural case' or, more precisely, a present type of extreme cultural colonialism. Any perspective in this sense cannot have a fixed framework in conceptual terms and, therefore, cannot be associated with any Western/European traditional or historical view. Any sense of 'independence' or 'liberation' is a dead end approach to the concept of Latin America as described here.

exploring the local dimension of the concept of environment we face a sort of discontinuity between understanding – based and conceived locally – and explaining, communicating and writing – where the local is forced to deal with the historiographic, epistemological and commercial Western/European tradition. Conceiving the local by interpreting something – in the hermeneutic sense, for example as Mignolo (2000) does – is not only useful but also possible in relation to a concept of Latin America. However, within that context, producing or generating cultural objects or artefacts that are able to establish a non-asymmetrical relationship with equivalents in the Western/European domain, is something that has proved to be much more difficult and somehow conceptually detached from any interpretation in the hermeneutic sense. And it is this discontinuity between interpreting and acknowledging, between interpreting as understanding and creating an object or theory as explaining, to put it in traditional terms (Wright, 1971), that will concern us in this chapter. In fact, exploring the connection between science and empire, as we did in Chapter 3, was the first general attempt to address this point.

3.1 Environment and 'Natural Selection'

The confusion between (i) a physical/material system, (ii) an idea of Nature, and (iii) a conception of fact (and evidence) is the most relevant aspect of a still dominant view/perspective of the meaning of 'world' when referring to Latin America. The reasons and arguments behind this convenient confusion between a physical/material system and an idea of Nature was largely explored elsewhere (see Canaparo, 2003: 829–893). What we would like to point out here is that the mechanism by which what is 'physical' is considered as natural, is the same mechanism that associates science with technology (see Canaparo, 2005b: 71 ff.), matter with reality (Canaparo, 2007: *passim*), and all that is considered 'factual' with a certain conception of truth and/or validation and legitimacy (Canaparo, 2003: 455–500).

What is relevant in the peripheral condition of Latin America, insofar as it is described here, is that all these conceptual associations and mechanisms get *corrupted*. They are not directly questioned on their argumentative status, which would be impossible from the conventional, commercial or intellectual point of view, they simply get twisted, and become implemented in an unexpected way. In this sense 'Latin America' is a place where, contrary to what some authors might think (see for example Larraín, 2005), the *implementation* of Western/European models and projects does happen, but in a *degenerative way*. The main paradoxical effect of Latin America, conceived as an environment and as described here, is that it produces a *de-naturalization* of Western/European ideas and perspectives. And this is exactly the limitation of Western/European colonialism: of course it exists, but not in a predictable way; it certainly has catastrophic and brutal consequences, but not in predictable way. Latin America is not a sociological laboratory of Western/European dominance, as it might appear after reading some local and non-local analysts. Of course, this is not because of any kind of 'resistance' prevailing locally, but quite the contrary. It does not happen for 'good' or positive reasons: it is because of the destructive and degraded nature of its environment and development. And this explains why the most successful forms of colonialism and/or imperialism are those that use political concepts to hide a systematic scheme of violence, corruption and State terrorism. Far from being a privilege of military governments or military regimes, this situation is present in many of today's formal democracies.

The worst intellectual attitude ever in Latin America is the one that tries to work with established *principles* of any kind, particularly those that focus on knowledge or ideas about knowledge. And that is why local authors like H. Maturana (1995a), when conceiving the notion of nature and *deriva* as a result and not as a point of departure, are quite effective and have therefore managed to engage themselves in a exchange – peer to peer – at an international level, instead of claiming that colonialism and imperialism are the main reasons for the lack of efficacy from the theoretical point of view (see for example the discussion about E. Dussel in Mignolo, 2000: chapter 1).

Paradoxically, in this context any proposal of 'liberation' or 'independence' paradoxically needs a consolidated Western/European paradigm *in order to be* what their authors claim the purpose of the theory is, leaving any attempt at 'originality' in a complete vacuum. On the contrary, from the perspectives discussed here, *more engagement* with the prevailing conceptual systems is the only way to ensure the construction of a local environment, the constitution of an *ad hoc* notion of knowledge, and an effective intellectual autonomy.

3.2 The Four Forms of Space

In line with the proposed forms of *environment* (Table 2.3), the three conceptions of *ground* (Table 1.2), and the aforementioned evolution of space itself (Figure 2.1), together with the elements and characteristics indicated in Chapters 1 and 2 concerning the constitution of a 'local space', it is now time to explore the idea of space itself more closely. Afterwards we will propose four definitions and/or forms of space as indicated in Table 4.1.

Table 4.1 shows the four definitions of space as considered in relation to Latin America considered as a concept or as a conceptual evolution. Obviously these forms are expressed in an 'estado puro' (pure state) but they normally appear linked. In other words, we never find one form of space operating 'purely' but rather two or three working together, with a certain characteristics from one form working together with characteristics from another.

Definition 1. Space is considered as *place*, that is, as establishing one *lugar*. More than determining by reference points, this space is defined by the characteristics of a particular and limited domain. And this space can be considered as a *place-world* when it is determined by, or in relation to, a position – in which case it is the context, the placement in a context that is relevant.

Definition 2. Space is determined by the *localidad*, that is, by a certain number of elements that make the nearby areas – the neighbourhood – clear and relevant. That is also why it is understood as *sitio* in some cases, i.e., as a particular and well-focused location.

Definition 3. The 'milieu' is a complex form of space that is produced by the interaction between a particular form of surrounding – or 'periferia inmediata' – and the sense of group obtained from its constitution by people or applied to objects ('community').

<table>
<tr><td rowspan="2">Ground stricto sensu</td><td>Place</td><td>'Lugar'</td><td rowspan="2">Definition 1</td><td rowspan="3">Question of location</td></tr>
<tr><td>Place-world</td><td>'Ubicación'</td></tr>
<tr><td rowspan="2">Cultural ground</td><td>Site</td><td>'Localidad'
'Sitio'</td><td>Definition 2</td></tr>
<tr><td>'Milieu'</td><td>Surroundings
'Periferia inmediata'
Community</td><td>Definition 3</td><td rowspan="2">Question of comprehension</td></tr>
<tr><td>Epistemological ground</td><td>Environment</td><td>Atmosphere
Weather
Forecast</td><td>Definition 4</td></tr>
</table>

TABLE 4.1. The four definitions of space that inform our comments on the conceptual construction of Latin America from the point of view of space. A comparison with the notions of *ground* (Table 1.2) and evolution (Figure 2.1).

Definition 4. The environment understood as space works here as an equivalent of *atmosphere*: space is conceived as a total set, that is, not only as space but also as all the elements included in it and, no less significantly, all the possibilities ('pronósticos', forecasts) that they can generate. This space has a close relation to the three forms outlined in Table 2.3, that is, with the ideas of 'ambiente', 'escenario' and 'amueblamiento'.

It is difficult to discuss a viewpoint based on the notion of space as regards the conceptual existence of Latin American without considering all these definitions. However, rather than discussing each one of them specifically, we have tried to implement them in our reflections and analyses, as we did previously with regard to the River Plate area (see Canaparo, 2005a and 2005b) and to the Patagonia area (see Canaparo, forthcoming). Less important than developing yet another taxonomy is the fact that certain definitions are better placed than others to more profoundly and precisely discuss a particular concept or conceptual evolution.

3.3 The horizon d'attente

Perception, understanding and the 'visiones de mundo' (world-view) are organized, according to the hypothesis proposed here, around the notion of *logistics*. As a set or ensemble, understood as 'knowledge', they are organized in *biological* terms (see for example Maturana, 1995a and 1995b). Of course, we say 'biological', not because the ultimate explanation about knowledge should *per forza* be reduced to biological descriptions, which is an ontological question/issue, but on the contrary, because biology understood as a cognitive process (see Maturana/Varela, 1980: 5–6) provides an accurate and holistic explanation of knowledge in terms of 'peripheral thinking' as we are discussing here.

It is when European authors such as P. Virilio, for example, try to consider of the limits of culture, society and understanding, that is when the relation with peripheral areas becomes more evident (see for example Canaparo, 2005b: 71–84 and 205–220). The 'logistics of perception', as Virilio himself referred to it, is an attempt to describe how far human perception is from the traditional European history of the definition of philosophy and how *artificial* it has become. The connection between war and cinematography is a privileged scenario that, according to Virilio, allows us to deconstruct this general artificial and non-reversible sense of our environment. This condition is pretty much applicable to the borders and peripheral areas of the Western/European domain. But this does not, as some simplistic extrapolation might suggest, account for the whole situation. And it can be understood by looking at the 'Europeism' that Virilio at no time tries to hide (see for example Virilio, 1996). The local use of apparatus, technological devices, the idea of reality, the concept of physical objects, *languaging*, the domain of things (physical and imaginary), are totally different from Western/European patterns. They function in a different way, their *functionality* is different, even when they work in a similar framework to the Western/European domain ('global market', 'scientific paradigm', 'culture pattern'). And it is this *functionality* – what Maturana, describing the epistemic dimension of a cognitive process based on biological parameters, calls 'deriva' (Maturana, 1995a: 109 ff.) – that makes the difference.

However, the group of perception, understanding and world-view, considered as a set/ensemble, as 'knowledge', is not only established around this functionality but also *originally* (*ab initio*) built up on each occasion and following a 'constructivist approach' (see for example Glasersfeld, 1995), as displayed in Figure 4.1. What we are trying to say is that, in terms of 'knowledge'/*languaging* understood as the ultimate expression of a particular community's major asset (Maturana, 2004), the immediate local implementation of the past and a particular observer or group of observers (see Glasersfeld, 1995: 113 ff.) are as important as the classical deciding past factors (traditions, colonialism, historiography).

Understanding then is the central element in what refers to the *horizon d'attente* situation. If we follow the classic Western/European philosophical taxonomical distinction that Georg Henrik von Wright (1916–2003) made between 'explanation' and 'understanding' (Wright, 1971), we end up with two intellectual traditions expressed in many ways: description and theory, prediction and explanation, Aristotle (384 BC–322 BC) and Galileo (1564–1642), causal methodology and teleological methodology, natural science and human science (see Wright, 1971: Chapter 1).

However the 'understanding' postulated here is far from the one associated with this Western/European interpretation of the history of philosophy. It is more closely related to the constructivist approach described by E. von Glasersfeld and J. Piaget (see for example Glasersfeld, 2005: 53–75; and Piaget, 1970a and 1970b) as a concept on its own and as the best connection with a perspective that situates language (*languaging*, construction of reality, construction of knowledge) at the heart of human activity. In this context, understanding is a comprehensive notion of four different situations: (i) 'we are someone', (ii) 'something is already there' (dictionary, encyclopaedia, physical world), (iii) 'we are looking at something' ('devenir', 'deriva'), and (iv) 'we need to take a stand, we need a particular direction'.

Understanding then, acting at the centre of the *horizon d'attente* attitude, works in three different directions: (i) from the *languaging* point of view exploring uncharted areas, (ii) from the knowledge point of view dealing with the perception understood as logistics and as vision, and (iii) from the physical point of view dealing with objects

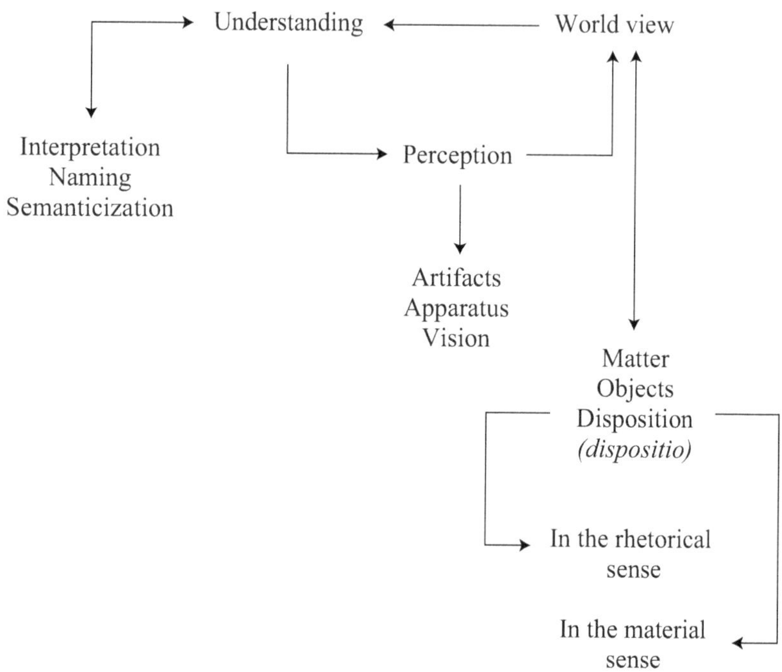

FIGURE 4.1 General outline of the mechanism of *horizon d'attente* in relation to the concept of Latin America defined by a spatial perspective.

and matters ('materiales'). Directions (ii) and (iii) also deal with the construction/modification of a 'world-view'.

Perception understood as vision and artefact, and the 'world' understood as a particular disposition of matter and objects, are the two main assumptions of the outline presented in relation to the operativity/operability of the *horizon d'attente*. In which case the present, the immediate environment and the thinking, are defined in relation to these assumptions.

And with this situation of the present understood as the horizon of expectation, the standardization of the way in which we construct our environment becomes more and more unified: 'after the standardization of opinion, which began in the nineteenth century, we are now seeing the sudden synchronization of emotions.' (Virilio, 2002a: 62).

This is the relevant point here: the *horizon d'attente* is relevant because, as we have already stated, it explains the construction of a particular form of *present*. It is relevant also because it determines the notion of 'knowledge' itself or, more precisely, because if our environment – including ourselves – is *a result* that cannot be approached in a retrospective way (Maturana, 1995a: 105 ff.), then the analysis is translated to the expectation of what the natural 'deriva' will make possible. In this sense, any concept that refers to Latin America, as analyzed here, is a device that works towards what is not yet here – and this working in a void or within what is not there, becomes one of the main characteristics of Latin America itself.

In Figure 4.1 we were trying to establish the basic conditions under which the *horizon d'attente* – understood as a main element of the present – operates. And if understanding is at the heart of this constructive process of knowledge/acknowledge it is because *languaging* – as described before – is the driving element of our experience:

> La experiencia humana está libre de contenido. En nuestra experiencia no nos encontramos con cosas, objetos, o naturaleza como entidades independientes como podría parecernos en la ingenuidad de la vida diaria; vivimos en la experiencia, en la práctica de vivir de seres humanos en el flujo de ser sistemas vivientes en el lenguaje como algo que nos sucede en nuestro interior y a nosotros mismos mientras lo lenguajeamos. (Maturana, 1995a: 95).

Furthermore:

> Los seres humanos existimos en el lenguaje, y nuestra experiencia como seres humanos se lleva a cabo en el lenguaje en un flujo de coordinaciones consensuales de coordinaciones consensuales de acciones que ponemos de manifiesto en el lenguaje. Los objetos, la consciencia, la autorreflexión, el ser, la naturaleza, la realidad, etc.; todo lo que los seres humanos hacemos y somos se lleva a cabo en el lenguaje porque usamos el lenguaje para generarlo, pero no como una abstracción o mero discurso, sino igual de concreto como cualquier operación en el flujo de las coordinaciones consensuales de acciones en las que surgimos y existimos. El hecho que existimos en el lenguaje, y constitutivamente no podemos existir fuera de éste porque estamos constituidos dentro de él, y el hecho que al ser en el lenguaje sólo generamos experiencias en el lenguaje, no nos limita, sino que al contrario, es la condición que hace posible que la ciencia sea un dominio explicativo en el que cualquier cosa que pongamos de manifiesto en él se convierte en parte de nuestra existencia como seres humanos. De hecho, el hecho que al existir en el lenguaje nuestro dominio de experiencia

> deba ser un dominio cerrado del cual no salimos, ni podemos salir, parece ser una limitación sólo si pensamos que deberíamos de poder referirnos a una realidad independiente. (Maturana, 1995a: 96)

This situation of experience is co-related to the notion of space as understood here and, more precisely, to the environment as something that has become a result and not a pre-condition of understanding.

Here is where we find a crucial point that we mentioned earlier in relation to Mignolo's views about what is Western/European and what is not. Colonialism, for example, does not feature in the arguments or in a particular subject as it normally happens in the classical way of understanding and defining arguments and subjects, that is, in an *explicit* form. And this perspective, the discussion of this perspective, is not an exclusive privilege of non-Western/European authors, and nor can the intellectual limitations for discussion of this perspective be defined by topological views. This is where we differ from Mignolo's implicit suggestion concerning the limits of European authors' ideas about, for example, colonialism (see Mignolo, 2000: 91–126). The moral or ethical views applied to historic evolutions of cultural analysis can only generate privilege and exclusivity (in the best cases) or new forms of colonialism (in the worst cases). The idea that the most adequate analysis of Latino culture, to take an example introduced by Mignolo, can only be produced by an author able to show some Latino 'credentials' seems as deplorable to me as believing that only homosexual authors can write about homosexuality. This is indeed part of the new colonialism – hiding behind some 'politically correct' statements – which comes from the US universities which try to establish the conceptual extension of Latin America.

And if we introduce this contentious subject here it is because we believe there is a close connection between this situation of what is possible or not possible according to a particular author's view and the extended domain within the Western/European domain – and everywhere – of the general attitude of *horizon d'attente*. The lack of acknowledgement of the consequences of the notion of 'present' by the *horizon d'attente* is one of the reasons why radical views about Western/European intellectual heritage can only take refuge in some utopian and illusionary desire for radical social change by anthropological acquisition. As is obvious to anyone living in Buenos Aires, São Paulo,

Santiago, Quito or Mexico, this is something that will never happen outside academia.

What is tragic about colonialism and, specifically, about the Western/European intellectual and cultural heritage in relation to Latin America is the fact that people – indians, natives, 'criollos' and migrants in general – constructed themselves around it. It therefore cannot be avoided, merely re-formulated in the best possible cases. Furthermore, this construction is strongest when it is aimed at the future, that is, towards what has not yet happened but which might happen. And it is in this future that, as we indicated before, the present itself is defined. It is this post-industrial mechanism of the formation of communities and individuals (Jeudy, 1990), well established in Latin America, that constitutes the real challenge to the Western/European tradition.

3.4 The Culture of Accident

It is in connection with this general state produced by what we describe as *horizon d'attente* that we find the second main aspect of knowledge. We will call this *the culture of the accident*. Furthermore, developments in previous chapters allow us now to go one step further and make clearer what was implicit from the beginning of this book: 'culture' understood in the Western/European conceptual tradition is not functional or effective in relation to what we call here 'Latin America'. As we stated in Chapter 2, even if we consider a holistic meaning of culture in conceptual terms, there is no 'natural' or 'historical sense' that could legitimize such a concept in Latin America. Culture, in the best possible way, can only be understood as *accident*.

Within the Western/European tradition the notion of accident has a long historiography, which starts with the Greek philosophers, gets a very important development during the Scholastic period and becomes even more specifically defined after the seventeenth century. For Aristotle, for example, the idea of accident was understood as συμβαίνει and was considered a philosophical notion to be explored on a logical and ontological bias. During the seventeenth century this field of implementation was extended to language and grammar. And even the Greeks raised a question about one 'science of the accidents', which was the question about the 'being' of the accident (κατά συμβεβηκός). The first systematic view about the accident is attributed to Porfirio (*circa* 233–305) in his *Isagoge*. But it was with the Scholastic authors, in relation to the philosophical discussions of their time, that the concept of accident developed fully. According to Léon Baudry, in William of Ockham (*circa* 1288–*circa* 1348), for example, there are more than 40 definitions of accident (see Ferrater Mora, 1994, I: 40). What is also interesting in the development of the Scholastics' views is the fact that the accident is considered not only in its content (substance) but also in itself (accident of the accident), from which the question of *accidenteïtas* ('accidentality') was developed. Paradoxically, this is one of the most relevant aspects that as we will see later on, P. Virilio developed in relation to contemporary Western European philosophy. From the seventeenth century a more detailed *grammatical definition* of accident was developed and, therefore, the logical and ontological perspectives were also modified.

Συμβαίνω in Greek has various meanings: (i) to have the feet (or boots) 'attached' (ensemble), (ii) get together with someone, (iii) come to terms with someone, (iv) agree on, and also has many meanings referring specifically to things like, for example, 'to let things drop' ('dejar caer'). The contemporary meaning of accident, more related to the Latin notion of *accidens/accidentis* – and in particular to *accido/is* – also has several significations: (i) fall, fall near to or fall over ('cadere presso, cadere sopra, cadere'), (ii) jump over ('piombare addosso'), (iii) 'come up to the ears' ('giungere all'orecchio'), (iv) becoming ('accadere, avvenire'), (v) manage to, obtain a result ('riuscire, avere un risultato'), (vi) cut ('tagliare'), (vii) to weaken ('indebolire, annientare'), (viii) to consume, to finish ('consumare'). All these different meanings are in the background every time we use the word accident just as they are when the notion of accident that we propose here is situated in a particular context. So accident, like other concepts, can be generally considered as a 'word' and at the same time as a 'notion', and it is the combination of these two contexts that produces the concept/s of accident.

As mentioned previously, if we understand culture as the most epistemically relevant set of characteristics, elements, bodies and objects of a particular space (see for example Knorr-Cetina, 1999), then the evolution of such a set, according to our perspective, can be understood only by *accident* and/or *accidentally*. And that is why the idea of change, understood as *accident*, is a concept that inevitably emerges when we refer to any form or aspect of knowledge. The concepts of evolution and modification that obtain nowadays on a conceptual, social and cultural level are closer to a notion of accident than to a traditional meaning of change:

1. Accidents happen randomly and can hardly ever be predicted.
2. Accidents represent a radical variation for the elements involved in them.
3. Accidents also cannot be predicted by their own internal development. Accidents work by 'ensayo y error' (trial and error).
4. Accidents are non-compatible with memory, or, to put it better, accidents scarcely have memory. However, there is a *know-how* of accidents and that is where the majority of the analyses concentrate.
5. Accidents are qualities or states that in principle are not part of the original structure, system, object, etc., that they affect or modify.
6. Accidents happen in language, in bodies, in things as in objects, and in individual people's mind.
7. Accidents cannot be defined by themselves but by what they produce or generate.
8. Accidents happen very quickly: an increasing speed is one of the regular aspects of accidents.

These characteristics of accident never appear clear or in sharp individual form. They appear more as an *evolution*, as H. Maturana described in his 'biology of cognition' (see Maturana, 1995a). The idea that evolution is an element that cannot be avoided when we are referring to knowledge is something that appeared very clearly in Maturana's theory (see for example Maturana, 1995a: 105 ff.). Furthermore, in Maturana's theory, the idea that evolution cannot be avoided is connected to the fact that things are not determined by any 'objective' condition:

> Desde un punto de vista biológico, los seres vivos son sistemas determinados estructuralmente. Por tanto, todo lo que les ocurre en cada instante les ocurre como parte de su dinámica estructura en ese instante, y determinado en ella. Ello implica que los cambios estructurales que un ser vivo sufre como consecuencia de sus interacciones con el medio en el que se realiza, no son determinados por los agentes externos que un observador ve incidir sobre él, ni dependen de la naturaleza de ellos, sino que quedan determinados por la propia dinámica structural del ser vivo (Maturana, 1995a: 108)

Moreover, according to Maturana

> el ser vivo no puede distinguir en su operar el origen interno o externo de sus cambios. En otras palabras, la dinámica structural de un ser vivo ocurre ajena a lo que un observador pueda distinguir como interno o externo a él. La distinción entre lo interno y lo externo, en un ser vivo o en cualquier otro sistema determinado en su estructura, es una distinción que hace el observador, y por esto no participa en el operar de tal clase de sistemas. De esto se sigue que los fenómenos propios de la dinámica estructural de un ser vivo y los fenómenos propios de su relación e interacción con el entorno en el que se realiza, son fenómenos de distinta clase y no pueden ser reducidos unos u otros. (Maturana, 1995a: 108–109)

Therefore, the link between one phenomenological domain ('sistema autopoiético', 'deriva') and the other ('ser humano', 'languaging', 'observador') is necessary and indispensable but also *accidental*: the environment of the 'ser vivo' ('living creature') cannot be detached from the specific 'ser vivo' that constitutes it, yet the determination of the environment itself is done by *accident*, by 'deriva', that is, by an exchange where structures and changes are established and made understandable by the change itself:

> …en un sentido estricto no hay ni puede haber determinación genética de ningún carácter o rasgo de la conducta de un organismo, pues éstos surgen de una manera sistemática durante la deriva ontogénica de éste. (Maturana, 1995a: 120)

The *accident* not only characterizes change in relation to its contents (first phenomenical/ phenomenological domain), it also refers to the way in which this change happens (second phenomenical/ phenomenological domain). The accident is related both to the *internal* and to the *external* dimensions of evolution, even when both domains according to Maturana are not directly related (see Maturana, 1995a: 118 ff.).

In the perspective postulated by Maturana the biological structure of human beings became an epistemic matrix for understanding understanding and for understanding how knowledge works. Therefore, the notion of 'deriva natural' and evolution are also regarded as entangled with this biological process and structure. In fact, the notion of 'biology of cognition', used to describe Maturana's works, is a clear indication of this entanglement. In fact, using Maturana's epistemic ideas to illustrate the notion of *accident* happens because they contain, in their characteristics – as described above – a clear symbiosis with Maturana's notion of evolution in terms of 'deriva natural': the 'deriva', as the evolution and the accident, are neither principles nor *a priori* but results, consequences.

In this sense, for W. G. Sebald (1944–2001) accidents are linked with the very nature of communities and the way they evolved. In his work *On the Natural History of Destruction* (2003) Sebald explores the entanglement between technology, understood as the natural history of our time, and destruction – a context in which accidents become the logic of the development and the change in human being. For different reasons *Vertigo* (1999) and *The Emigrants* (1996) both tend towards a point where speed and movement get entangled with displacement and migration, always in a natural context given by technology and post-industrial landscape.[16]

What is really instructive in Sebald's works is the way in which he incorporates war environments, technology, destruction, media and speed to create a general view about society and culture. And in this sense Sebald's work can be related to Robert Bevan's *The Destruction of Memory* (2006) as we have mentioned elsewhere (see Canaparo, forthcoming). There is a close relation between destruction, war environments, technology and the construction of perception, not only with reference to a possible common set of elements in a particular community – that we can call 'culture' – but also, and sometimes more relevantly, in relation to the establishment of a notion of present and

16 According to Virilio, 'In the future, just as the geographic world was colonized by means of transportation or communication, we will have the possibility of a colonization of the human body by technology' (Virilio, 1997: 6).

reality. This is also the clear message in Sebald's works – a hypothesis that places him among such European thinkers as J. Derrida and P. Virilio (see for example Derrida, 1972: 365 ff.; and Virilio, 1989).

Following on from this semantic and instrumental connection between accident and war it is interesting to briefly explore one or two works by Ernst Jünger (1895–1998). For Jünger 'war' is neither a situation nor a historic period but a *full environment*. And in these circumstances, the environment is a product of *accidents* which generate irreversible changes.

The magnificent *In Stahlgewittern* (*Storm of Steel*, 1920) makes exactly this point: what happens in war is not so much a matter of rationality or strategy as a *summa* of accidents.[17] Within a war environment almost everything in the battlefield (and outside it) is defined by accident,[18] and in a war environment all space is a battlefield.[19] This perspective is in line with some of the classic works about war in the Western/European tradition like *Im Westen nichts Neues* (*All Quiet on the Western Front*) (1920) by Erich Maria Remarque (1898–1970), and have a direct relation with the creation of a 'cultural space' in places like the River Plate area, as we have discussed already elsewhere (see Canaparo, 2005b: 121–147).

The de-composition and displacement of individuals and bodies exhibited in *Die Totale Mobilmachung* (1930) and *Feuer und Bewegung* (1930), is not detached from the implosion of communities and architectural systems shown in the collective work *Guerra, técnica*

17 As is well known, the first version of *Storm of Steel* was published in the author's private edition in 1920. Since then several editions and modifications were produced by Jünger, with the 1961 edition being considered as the final one, made for the *Collected Works*. The main idea referred to here remains at the centre of all the versions.

18 Moreover, an idea that appears continually in Jünger's works – as it does in other authors, like Louis Ferdinand Céline (1894–1961) – is that in a war environment the battlefield becomes the paradigm of space: everything becomes defined by relation to it.

19 As we indicated before in this chapter, P. Virilio would postulate the hypothesis that our contemporary social imaginaries were generated from war environments and that somehow we constantly live – due to technology, construction of our perception and use of the media – in a war environment (see Virilio, 1984).

y fotografía (2002). In Jünger's works 'war' and accident are two aspects of a general phenomenon that can only be expressed (naming, visualizing) by the concept of environment.

Following this *déroute*, for Paul Virilio, perhaps more than for any other contemporary European philosopher, *accidents* are the substance of our present knowledge (see Virilio, 1997: 7). In *Unknown Quantity* (2002a) as in *Ce qui arrive* (2002b), many perspectives on this main direction are explored. However, what interests us here is Virilio's proposal that it is no longer possible to make a distinction between what can happen (future, *ce qui arrive*) and what is happening now (present, *devenir*) (see for example Virilio, 1993 and 1996): everything seems to be happening all the time ('ce qui arrive'). The future as potential present becomes as real as the perception of the present itself because, to construct this perception of the present, perspectives about what might happen are fundamental (Virilio, 1988 and 2002a). Moreover, from the point of view of a contemporary notion of knowledge all elements and perspectives are dominated by a *horizon d'attente* (horizon of expectation) as described here (see above in this chapter).

It is also in this context that the notion of change understood as *accident* became particularly relevant and radical, because any possible notion of present – as a phenomenon that is happening now – became a domain completely controlled by a 'horizon of expectation' (see for example Jeudy, 1990). The idea of change cannot be considered outside the definition of the present itself – as 'devenir' (turn into) or pure becoming –, therefore all 'social' or 'cultural' issues – as main elements regarding a particular community of group – can only be referred to (named, indicated, acquiring meaning) by their relation to a perception constructed around the 'horizon of expectation'. That is why Virilio said that 'the perception of accidents is always an "accident of perception"' (Virilio, 2002a: 56–57). And if Virilio had a relation with this 'culture of the accident' as we are here describing Latin America conceptually it is because, as a European philosopher, he went to the extreme of our contemporary culture. He did this by indicating that there is a close relation between (i) the way we construct our perception as 'an apparatus', (ii) the way in which technologies have been developed in post-industrial societies, and (iii) the way how violence has been re-formulated as an unavoidable catastrophe or natural destruction.

Reality for Virilio, in a similar way to the concept of Latin America that we are discussing here, became more and more an accidental phenomena (Virilio, 2002a: 87). Thus understanding then became an 'accident of knowledge' (2002a: 7) and the immediate environment understood as present became a constant threat expressed by the 'drunken driver's words to his passenger: "I'm an accident looking for somewhere to happen"' (Virilio, 2002a: 9). And, for precisely this reason, knowledge understood in terms of 'accident' cannot be taxonomized:

> 'There is no science of the accident', warned Aristotle long ago. In spite of *cyndinics*, the new science of risk-assessment, there is no *accidentology*, but chance discovery, archeo-technological invention: to invent the sailing vessel or the steam ship I to *invent the shipwreck*. To invent the train is to *invent the derailment*. To invent the private car is to produce *the motorway pile-up*. To make craft which are heavier than air fly – the aeroplane, but also the dirigible – is to invent the *plane crash*, the air disaster. (Virilio, 2002a: 23–24)

Moreover, this impossibility of predicting or rationalizing is linked directly to the fundamental role of the 'horizon of expectation' already discussed: when the idea of present is determined by expectations about what can happen, it is the status and meaning of this 'happen' (or 'happening') that also changes.

However, contrary to what Virilio maintained, we do not believe it is possible to make a distinction between 'natural' and 'man-made' accidents (see Virilio, 2002a: 23): all accidents are as natural as man-made. The comprehension of 'natural' accidents requires an equal amount of technical devices as the comprehension of 'man-made' ones. It is the unavoidable technological dimension of our environment that makes any concept both artificial and natural at the time: 'artificial', because it is generated and/or understood by devices and apparatus, 'natural', because (as before) it forms our immediate environment, what Maturana calls *nicho* (niche) (Maturana, 1995a: 111). The irreversible industrial and technological condition of our environment is what makes accidents the most natural form of change (see for example Virilio, 2002a: 23–24). Today we are in the absolute opposite situation, as described by the French humanist André Malraux (1901–1976) sixty years ago:

> Culture is what makes man something other than an accident of the Universe. (see Virilio, 2002a: 27) [20]

We are at the exact antipodes of Maulraux's *dictum*. Malraux's conceptual man no longer exists, except as an anthropological event of the past. We are living in an environment built up around the culture of accident. Such a 'culture' is not anthropomorphical anymore, but is a 'culture' that is an *unknown quantity* of aspects, concepts, things, objects, all of them related to a present – or, more precisely, to a set of actual contemporary events – by *accident*. And in this context, what is the meaning/concept of language, historiography and established physical systems, is less related to the definition of 'culture' than to how they – as a part or a totality – are connected to that notion of present.

The understanding of culture in terms of accident reinforces the *meteorological* dimension of it, the fact that culture behaves also in terms of weather: regular forecast, instabilities, predictions, sense of the local and the global, sense of short term and long term. In the words of Virilio:

> Just as there are spells of bad weather in nature, there are also periods of bad weather in culture and we would need a positive 'meteorology' of invention to attempt to avoid *the storms of the artifice of the Progress of knowledge*. (Virilio, 2002a: 86)

As we have already discussed in Chapter 1, the fact that we are continually forced to think – and define knowledge – in terms of atmosphere, gives this meteorological perspective an everyday quality. Not only are we looking every morning for the traditional indicators of temperature, raining, etc., we also look at the indicators of pollution, CO_2, decibels, etc., and we dress and carry devices accordingly. More relevantly: we cannot understand our physical surroundings other than through this atmospheric perspective. And this is why Maturana's

20 A more complete quote from Malraux states that culture 'would seem, first and foremost, to be the knowledge of what makes man something other than an accident of the universe, be it by deepening his harmony with the world or by the lucid consciousness of his revolt from it.' (see André Malraux, *Appreciating Coincidences* at www. scholar.lib.vt.edu/theses/).

notion of the biology of cognition is so useful: because it allows us to think about this atmospheric condition as our *natural environment*.

In these *accidental* conditions, the movement of bodies and objects and, principally, the travelling of bodies and ideas, become transformed in displacement and physical translation. It is no longer the people or their bodies that are nomadic but the geography, the landscape, the land, the ground. This is exactly the experience described regarding one of the most extreme forms of nomadism: the so-called transient mental illness or, more specifically, 'mad travellers'.

Ian Hacking (1936–) wrote a very interesting book on precisely this subject entitled *Mad Travellers: Reflections on the Reality of Transient Mental Illnesses* (1998). At first glance, what I found most interesting in it is the fact that 'nomad' and 'fugitive' are associated elements from the beginning, which certainly follows the first works on the subject: *Les aliénés voyageurs* (1887) by Phillipe Tissié (1852–1935) and *Les aliénés voyageurs, ou migrateurs* (1875) by Achille Louis Foville (1799–1878). Another pertinent point here is the fact that, according to Hacking, the main purpose of these voyageurs' or fugitives' trips, was 'less a voyage of self-discovery than an attempt to eliminate self' (Hacking, 1998: 30). And these *dromomanies* (travelling fugues) as Etienne Regis (1860–1927) called them around 1893 once considered as 'anomaly', have today become of renewed interest as they are entangled and share similar characteristics with many massive collective phenomena, such migration, diaspora behaviour, war displacement, etc.[21] Contrary to the leading case of Jean-Albert Dadas studied by Tissié and discussed by Hacking,[22] nowadays the *dromomanies* are part of our culture or, more exactly, are part of how we perceive the physical world and, therefore,

21 'Dromomania', also called *travelling fugue*, is described by authors as a psychological condition in which people spontaneously depart from their routine, travel long distances and take up different identities and occupations. Normally months might pass before they return to their former identities.

22 Jean-Albert Dadas, was a gas-fitter from Bordeaux. Dadas would suddenly start travelling, on foot, and go far away as Prague, Vienna or even Moscow, with no memory of his travels. A medical student at the time, Philippe Tissie, wrote about Dadas in his doctoral dissertation in 1887, and this is now considered as one of the pioneering works in the subject.

how we form our knowledge of it. Maybe the term of 'automatisme ambulatoire', coined by Jean Martin Charcot (1825–1893), to deal with a similar case, is more appropriate to the understanding of Western/European analysts or authors – the avant-garde work of Virilio itself, *Vitesse et politique*, in 1977, and the influence it has had on many authors, is a clear confirmation of this. Furthermore, there is a sort of alienation of the *maladie* itself and the departure from routines – the indisputable symptom of it as described in the nineteenth century – can also take non-physical forms or, more precisely, can take the form of *motricité* where the fugue is done within a city or urban environment and not over long distances in the traditional sense. The notion of *picnolepsie*, explored by P. Virilio in the *Esthétique de la disparition* (1989: Chapter 1), can be considered as a contemporary form of this alienation as it is described and associated with forms of epilepsy – as is *dromomanie* – , absence of memory and regular absenteeism (Virilio, 1989: 11 ff.).

The constant state of *dromomanie* is the result of a culture of accident as we have described here or, to be more precise, it is the construction of the self carried out in this culture of accident that results in the implementation of random and unstable behaviours – like the *dromomanies*. Furthermore, as already mentioned, this *dromomanie condition* does not only refer – as in Tissié's case study – to something that necessarily has a physical and clear component (the departure of the body, the fugue). In the domain of knowledge, as approached here, the *dromomanie* can also be only *imaginary or mental*, in which case it can also be identified as *poriomanie* (πορεία) (see for example Solms, 1995 and Hallen, 1957).[23] However, what is radically different from the nineteenth-century concept is the fact that, if the notion of *errance* was previously defined by rapport to the *motricité* ('la mise en mouvement'), nowadays the *errance* is much more abstract as the notion of movement itself has become abstract and no longer has a *physical expression* that can be displayed from the conceptual point of view. The *errance* is still pretty much considered as something 'poussé… a' (move to, force to,

23 *Poriomanie*, from the Greek to 'walk' or to 'wander' (*marcha*), was considered an epileptic equivalent to fugue, without involvement of the consciousness, by the German scientist Donath in 1899.

push to) but this 'towards' does not always have a sense of physical translation even if it does always have a sense of displacement.

Finally what is really interesting about the *dromomanie condition* is that, even when Hacking does not discuss the issue specifically, many authors have agreed to strongly associate it with the development of *fantômes* (see for example Soundaramourty, 2003). They do so in such a way that perceiving, travelling itself and acknowledging places and people has become directly associated with the notion of *fantômes*. And so it is here, in the development of the culture of accident as part of our ordinary practices, that the notion of *fantôme* appears, (i) not only as a psychoanalytical term that may explain to the *dromomanies* (see for example Bolzinger, 2003), but also (ii) as an element that is present at the very core of an accelerated evolution and deployment of technology, the accidental situation described before, and, no less importantly, (iii) as a philosophical issue that is placed at the very heart of Western/European tradition (see Garré, 2004; and Caro, 2005).

According to P. Fuentes and M. Veli (2006), there is in S. Freud's works a close relation between narration in terms of *fabula* – literature considered as a hermeneutic tool – and the construction of a *fantôme*.[24] Furthermore, it would seem that the basic construction of the self – or image of the self – as a narrative goes hand in hand with some kind

24 'En "La novela familiar de los neuróticos" (1908–1909), Freud describe una actividad fantaseadora del niño que, en la prepubertad, se apodera del tema de las relaciones familiares. En estas fantasías noveladas suele incluirse una idea de sustitución de ambos padres o de uno de ellos por otros, habitualmente "personas más grandiosas" y "nobles". Freud observa que esta sustitución, más que eliminar al padre lo enaltece, ya que la figura que lo sustituye aparece dotada de rasgos provenientes del verdadero. La novela familiar viene entonces al lugar de la falla estructural del padre e intenta repararla. [...] En "Fantasías histéricas y su relación con la bisexualidad" (1908), Freud se ocupa de los sueños diurnos y dice que "se los cultiva con esmero y la más de las veces se los reserva con vergüenza". Este texto es el más claro antecedente freudiano de la conceptualización lacaniana del fantasma. [...]. Como señalara Roberto Harari en *¿De qué trata la clínica lacaniana?*, Freud califica al fantasma como novela, diferenciando entre *roman* y *nouvelle*, es decir entre novela y novela breve. El fantasma adopta la estricta condición de la novela breve, donde se representa y relata en forma abreviada un acontecimiento.' (Fuentes/Veli, 2006)

of 'phantasmatic elaboration'. And these phantasmatic tendencies have many collective formulations in Freud's works, such as the tendency towards self-destruction, as conjectured in his last works about culture and also in his *Moses* (see Freud 1930 and 1939), among the most notorious examples. Furthermore, as analysed by Federico Caro (2005), there seems to be a close connection between this 'phantasmagoric elaboration' and the *voyage pathologique*, in the sense that 'voyage' and *fantôme* are two aspects of the same equation in Western/European tradition (see Caro, 2005).

The relation between the *fantôme* and the fantasy in terms of psychoanalysis is also what led Freud to explore the idea of *fantôme* in an autonomous sense, as a concept.[25] Since then many works have been written on the subject, in particular those concerning the neo-Freudian theories of Jacques Lacan (1901–1981).[26] In psychoanalysis the *fantôme* is an 'image' that is not part of the real and that in some cases is established as a defence for certain neurosis or, in a more recent and less traditional interpretation, the *fantôme* is considered as an inevitable part of the psyche in the construction of the sense of reality. Therefore, it is not only relevant to identify and circumscribe the *fantôme*, as was done in the traditional psychology, it is also relevant to explore and develop its functionality.[27] Lacan's theory starts from the point of view that the *fantôme* is part of Western/European culture and constitutes an illusion

25 Freud was certainly aware of the double Greek meaning of 'phantasy', that is, 'hacer aparecer algo, que puede ser idea o imagen' and 'hacer nacer o surgir una idea, imaginación o representación', 'figurarse algo', 'representarse algo'. Both meanings acquire relevance with the concept of *fantôme*, in particular to distinguish it from the idea of 'imagination'. There is also a sense of *catalepsy* within *fantôme* ('phantasmata') that also becomes relevant.

26 With regard to the notion of *fantôme* within psychoanalysis see for example Nasio, Juan David. *Cinco lecciones sobre la teoría de Jacques Lacan* (Barcelona: Gedisa, 1993: 147–172). Also Jacques Lacan's *Le Seminaire*, Livre XIV, 'La logique du fantôme' (see Lacan, 1967).

27 'En el Seminario XIV, "La lógica del fantasma" (clase del 21 de junio de 1967), Jacques Lacan lo ubica alrededor del concepto lógico matemático de axioma, atribuyéndole una "significación de verdad". Dice también que consiste en una frase en la cadena de enunciados. El fantasma es cuando se dice. El axioma es una construcción a posteriori que no es ni verdadera ni falsa.' (Fuentes/Veli, 2006).

in the belief that it could be totally eradicated (see Lacan, 1967; and Milner, 1995).

And this perception of the *fantôme* is not something that has come out of nowhere. At the core of Western/European philosophy, anchored in René Descartes' works, there is this discussion about the commensurable/incommensurable dualism between mind and body. According to the British philosopher Gilbert Ryle (1900–1976) this unresolved dualism is what characterizes Descartes' tradition ('doctrine'). It was not by chance that Ryle identified this perspective with the expression 'the ghost in the machine', as used in his book *The Concept of Mind*, published in 1949.[28] Arthur Koestler's (1905–1983) famous work *The Ghost in the Machine* (1967) follows this interpretation and makes it more extreme. Koestler's view is that primitive structures of the mind remain in the human mind and randomly emerge with the worst results imaginable, particularly concerning what is referred to as self-destructive tendencies – tendencies that Freud also generally addressed even if for purposes different to Koestler's.

Paradoxically, what Ryle did not understand – as Koestler did – was the fact that the 'machinery perspective' of human being and nature was anchored in the very notions of science, progress and knowledge, framed by Descartes' philosophy and considered as the core of Western/European tradition. It is precisely this scientific sense of 'ghost in the

28 In *The Concept of Mind* (1949), Ryle admits to having been taken in by the body-mind dualism which permeates Western philosophy, and claims that the idea of Mind as an independent entity, inhabiting and governing the body, should be rejected as a redundant piece of literalism carried over from the era before the biological sciences were established. The proper function of Mind-body language, he suggests, is to describe how higher organisms such as humans demonstrate resourcefulness, strategy, the ability to abstract and hypothesize and so on from the evidence of their behavior. He attacks the idea of seventeenth- and eighteenth-century thinkers (such as Descartes and La Mettrie [1709–1751]) that nature is a complex machine, and that human nature is a smaller machine with a 'ghost' in it to account for intelligence, spontaneity and other such human qualities. While mental vocabulary plays an important role in describing and explaining human behavior, humans are neither analogous to machines nor do philosophers need a 'hidden' principle to explain their super-mechanical capacities.

machine' that has been developed following Koestler's critical work. The music album *Ghost in the Machine*, created by the British rock group The Police in 1981, is an astonishing example of this notion where technology and post-industrial societies are grounded in an accidental way, even when presented in a rational and well-organized manner. The sci-fi movie *Ghost in the Machine* (1993) by Rachel Talalay can be considered in the same way. Within the film, the notion of *fantôme* becomes a horrific one and it is difficult to establish the distinction between *fantôme* and nightmare (the Freudian perspective appearing again here). This is a situation that is consistent with Koestler's prediction about the '*fantôme* of the Western/European civilization'.

Henri-Pierre Jeudy's (1944–) works are also aware of this *déroute*, particularly with regard to the fact that, within this accidental condition of knowledge, the present becomes a construction made in the future or, to be more precise, in the expectation of what we believe can/might happen. One of his main hypotheses then refers to the idea that the determination of the present – and the 'actions' attached to it – is something that nowadays depends almost completely on the *horizon d'attente* (horizon of expectation), or, more exactly, on the expectations and networks of belief generated by it. This hypothesis was already studied by Jeudy in the European environment in his *Le désir de catastrophe*, published in 1990. What Jeudy also exposes is the fact that there seems to be a direct link between the construction of a particular environment and the exercising of a *horizon d'attente*. More specifically, there seems to be a connection between the way we make an environment *familial* and the way we determine the present via a *horizon d'attente*.[29]

The idea of a 'culture du risque, culture du désastre' identified by Jeudy (1990: 91–96) has similar grounds to the culture of accident studied here. We mention this link to try to show how *close* the epistemological issues are nowadays when we start to talk about construction of an environment, despite the radical differences that might exist between

29 A similar perspective is also followed by Marc Augé in *Où est passé l'avenir?* (2008).

a Western/European conception and a Latin American approach. The introduction of the catastrophic element into ordinary life, as something that might happen frequently, is something that destroys frontiers and traditions. It is a paradox that the more we try to establish and determine a notion of environment, the more we are exposed nowadays to a 'planet effect', that is to say, to an assertion that local roots and practices can be erased from one day to the next.

In *Fictions théoriques* (2003) Jeudy discusses some of the epistemological issues that emerged in *Le désir de catastrophe*. What interests us in particular is the proposed entanglement between speculation – viewed as written theory – and knowledge. Jeudy seems to think that theories – understood as *fictional tools* – cannot stand in a systemic way, that is, cannot stand (as was traditional) in a general-universal theory, because their reliability and existence can disappear from one day to the next without any rational explanation nor any explanation based on a common *point de repère*. 'To think' in this context is very close to the notion of thinking we have been discussing in relation to the peripheral areas of Western/European culture and traditions (see Chapter 1).

This brief *parcour* (journey) through the notion of accident in Western/European tradition, and the considerations about some of its developments, enables us to obtain a more specific dimension of the meaning of the 'culture of accident'. However, classical Western/European interpretations of culture try to deal with the extreme contemporary conditions of knowledge by appealing to a hermeneutic/phenomenological approach that – as the last big Western/European philosophical system – uses *time* (events, history, etc.) as the core element of understanding. One of the best and most interesting recent examples is Nicole Auber's (1957–) work *Le culte de la urgence* (2003).

However, what is relevant for us is the fact that, as stated, the culture of accident functions (i) as a *result* or (ii) as a condition of knowledge. And in these circumstances the culture of accident has three major effects on the construction of our immediate environment and knowledge:

1. The construction of the local. It functions by creating and developing a landscape perspective as a regular and ordinary tool for understanding our surroundings and 'vital space'. In this sense 'the culture of accident' leads to the generalization of a *manufactured landscape* as part of the immediacy and starting point of the local environment. The notorious documentary *Manufactured Landscapes* by the Canadian filmmaker Jennifer Baichwal (1960–) (2006) is a wonderful illustration of this point.
2. Learning/acting in a war environment. The war environment is the nature of ourselves, of our bodies, and of our knowledge. Evolution understood in H. Maturana's terms has this cultural/collective consequence and is regarded as irreversible. And this means not only that we have conceived 'life as war' – to use the expression from Godfrey Reggio's (1940–) film *Naqoyqatsi* (2002) – but also, and principally, that we understand and develop our environment as in/within war conditions. This is a crucial distinction between Western/European tradition and the Latin American perspective.
3. The making of knowledge. In this sense 'the culture of accident' leads to a *tiranía* (dictatorship) of the numerical aspects of capitalism and post-industrial societies (and the consequent development of concepts like 'quantity' or/and quantification). According to the statistical perspective that dominates our sense of *measure* in collective/societal terms, we die in advance nowadays: we are always part of a group of people that will die for something that is or has become *common*. As Norbet Elias (1897–1990) has already stated, the abstract aspect of our 'dying condition' is the worst psychological enemy of our intellectual and emotional integrity (see Elias, 1985).

Less because it was founded on folkloric independence, the concept of Latin America became effective because it offered a combination of (i) different forms of space (see Figure 4.1), (ii) a *natural* perspective based on a *horizon d'attente*, and (iii) a series of *accidental* aspects that determine our notions of environment and knowledge.

4. Bibliography

Aubert, Nicole. *Le culte de l'urgence* (Paris: Flammarion, 2003).
Augé, Marc. *Le Métier d'anthropologue* (Paris: Galilée, 2006).
———. *Où est passé l'avenir?* (Paris: Panama, 2008).
Barnes, Andrew. 'Review on *The Darker Side of the Renaissance*', in *Journal of Social History*, Spring, 1998.
Baudrillard, Jean. *Simulacres et simulation* (Paris: Galilée, 1981).
Benjamin, Andrew (ed.). *The Problems of Modernity: Adorno and Benjamin* (London: Routledge, 1991).
Benjamin, Walter. *Charles Baudelaire: A Lyric Poet in the Era of High Capitalism* (London: Verso, 1983).
Bevan, Robert. *The Destruction of Memory* (London: Reaktion, 2006).
Blaut, J. M. *The Colonizer's Model of the World. Geographical Diffusionism and Eurocentric History* (New York: Guilford Press, 1993).
Bolzinger, André. 'Folies cosmopolites', in *L'évolution Psychatrique*, volume 68, issue 3, 2003, pp. 447–455.
Bourdieu, Pierre. *Homo academicus* (Paris: Minuit, 1984).
———. *L'ontologie politique de Martin Heidegger* (Paris: Minuit, 1988).
———. *Les règles de l'art* (Paris: Seuil, 1992).
Canaparo, Claudio. *Ciencia y escritura* (Buenos Aires: Zibaldone, 2003).
———. 'The *Nature* Effect in Latin America Science Publications: The Case of the Journal *Redes*' in E. Ortiz/E. Fishburn, eds., *Science and the Creative Imagination* (London: Institute for the Studies of Americas, 2005), pp. 97–118. [2005a]
———. *Muerte y transfiguración de la cultura rioplatense* (Buenos Aires: Zibaldone, 2005). [2005b]
———. 'La consumación del realismo', in G. Fabry/C. Canaparo, eds., *El enigma de lo real. Las fronteras del realismo en la narrativa del siglo XX* (Bern: Peter Lang, 2007), pp. 199–275.
———. *El imaginario Patagonia. Ensayo acerca de la evolución conceptual del espacio* (forthcoming).
Cañizarez-Esguerra, Jorge. *How to Write the History of the New World* (Stanford: Stanford University Press, 2001).
Caro, Federico A. *Voyage pathologique: Historique et diagnostique differentiels*, Memoire du DEA, Université Paris 7, 2005.
Casanova, Pascale. *La République Mondiale des Lettres* (Paris: Seuil, 1999).
Cassirer, Ernst. *Le problème de la connaissance dans la philosophie et la science des temps modernes* (Paris: Cerf, 2004).
Cía Lamana, Domingo. *Narración y pensamiento. Hacia un Nuevo paradigma del saber* (Barcelona: Erasmus Ediciones, 2006).
Cusset, François. *French Theory* (Paris: La Découverte, 2003).
Derrida, Jacques. *De la grammatologie* (Paris: Minuit, 1967).
———. *Marges de la Philosophie* (Paris: Minuit, 1972).

———. *Mal d'archive* (Paris: Galilée, 1995).
———. *Le monolinguisme de l'autre* (Paris: Galilée, 1996).
Dussel, Enrique. 'Beyond Eurocentrism: The World-System and the Limits of Modernity', in F. Jameson/M. Miyoshi (eds.). *The Cultures of Globalization* (Durham: Duke University Press, 1988), pp. 3–30.
Elias, Norbert. *The Loneliness of the Dying* (Oxford: Blackwell, 1985).
Fabry, Geneviève/Canaparo, C. (eds.). *El enigma de lo real. Las fronteras del realismo en la narrativa del siglo XX* (Bern: Peter Lang, 2007).
Ferrater Mora, J. *Diccionario de filosofía* (Barcelona: Ariel, 1994), Volume I.
Foucault, Michel. *Les mots et les choses* (Paris: Gallimard, 1966).
Freud, Sigmund. *Civilization and Its Discontents* (London: Hogarth, 1930).
———. *Moses and Monotheism* (London: Hogarth, 1939).
Frisby, David. *Fragments of Modernity: Theories of Modernity in the Work of Simmel, Kracauer, and Benjamin* (Cambridge, Mass.: MIT Press, 1986).
Fuentes, Carlos. *La nueva novela hispanoamericana* (Mexico: Joaquín Mortiz, 1969).
———. 'Unidad y diversidad del español', in http://www.congresosdelalengua.es/rosario/inauguracion/default.htm, 2004.
Fuentes, P./Veli, M. 'Novela, mito y fantasma en los relatos de acientes de psicoanálisis', in *Psikeba. Revista de pisocoanálisis y estudios culturales*, number 1, Enero-Abril 2006.
Galeano, Eduardo. *Memory of Fire 2* (London: Quartet, 1986).
———. *Memory of Fire 1* (London: Meuthen, 1987). [1987a]
———. *Memory of Fire 3* (London: Quartet, 1987). [1987b]
Garré, J. B. 'Psychiatrie du voyage, voyages de psichiatres, voyage en psychiatrie', in *Actes de la XIXème Journée de psychiatrie de Fontevraud*, 2004.
Geary, Patrick. *The Myth of Nations: The Medieval Origins of Europe* (Princeton: Princeton University Press, 2003).
Gimeno Blay, Francisco M. *Escribir, reinar* (Madrid: Abada Editores, 2006).
Glasersfeld, Ernst von. *Radical Constructivism. A Way of Knowing and Learning.* (London: Falmer Press, 1995).
———. 'Distinguishing the Observer: An Attempt at Interpreting Maturana', in http://www.oikos.org/vonobserv.htm, 1997.
González Casanova, Pablo. *Estado y sociedad en América Latina.* (Mexico: Oasis, 1984).
González Echevarría, Roberto. *Myth and Archive. A Theory of Latin American Narrative* (Cambridge: Cambridge University Press, 1990).
Grosfoguel, R./Cervantes-Rodríguez, M. (eds.). *The Modern/colonial capitalist World-system in the Twentieth Century* (Westport: Greenwood Press, 2002).
Hacking, Ian. *Mad Travellers. Reflections on the Reality of Transient mental Illnesses* (London: Free Association Books, 1998).
Haller, W. von. 'The Problem of Poriomania', in *Nervenarzt.*, 28: 9 (1957), pp. 385–9, 1957.
Jeudy, Henri-Pierre. *Le désir de catastrophe* (Paris: Aubier, 1990).
———. *Fictions théoriques* (Paris: Manifeste/Editions Léo Scheer, 2003).

Jünger, Ernst. *Sobre el dolor* (Barcelona: Tusquets, 1995).

———. *Guerra, técnica y fotografía* (Valencia: Universidad de Valencia, 2002).

King, John (ed.). *The Cambridge Companion to Modern Latin American Culture* (Cambridge: Cambridge University Press, 2004).

Knorr-Cetina, Karin. *Epistemic Cultures. How the Sciences Make Knowledge* (Cambridge, Mass.: Harvard University Press, 1999).

Koselleck, Reinhart. *Futuro passato* (Rome: Marietti, 1986).

Krige, John/Barth, Kai-Henrik (eds.). 'Global Power Knowledge', *Osiris*, volume 21, 2006.

Lacan, Jacques. *Séminaire XIV La logique du fantasme 1966–1967.* (Unpublished typescript, 1967).

Larrain, Jorge. *Identity and Modernity in Latin America* (Cambridge: Polity Press, 2005).

Latour, Bruno. *Science in Action* (Cambridge, Mass.: Harvard University Press, 1987).

———. *Petites Leçons de sociologie des sciences* (Paris: La Découverte, 1993).

Lebeau, André. *L'engranage de la technique* (Paris: Gallimard, 2005).

Lakatos, Imre. *The Methodology of Scientific Research Programmes* (Cambridge: Cambridge University Press, 1978)

Lipovetsky, Gilles. *La société de deception* (Paris: Textuel, 2006).

Lipovetsky, Gilles/Charles, Sábastien. *Les Temps hypermodernes* (Paris: Grasset, 2004).

Maturana, Humberto. *La realidad: ¿Objetiva o construida?. I. Los fundamentos biológicos de la realidad* (Barcelona: Iberoamericana/Anthropos, 1995). [1995a]

———. *La realidad: ¿Objetiva o construida?. II. Los fundamentos biológicos del conocimiento* (Barcelona: Iberoamericana/Anthropos, 1995). [1995b]

———. *Transformación en la convivencia* (Santiago: J. C. Sáez, 2004).

Maturana, H./Varela, F. *Autopiesis and Cognition* (Dordrecht: Kluwer, 1980).

Mignolo, Walter. 'Signs and Their Transmission: The Question of the Book in the New World', in Hill Boone, Elizabeth/Mignolo.W. D. (eds.), *Writing Without Words* (Durham: Duke University Press, 1994), pp. 220–270. [1994a]

———. 'Afterword: Writing and Recorded Knowledge in Colonial and Poscolonial Situations', in Hill Boone, Elizabeth/Mignolo.W. D. (eds.), *Writing Without Words* (Durham: Duke University Press, 1994), pp. 292–310. [1994b]

———. *The Darker Side of the Renaissance. Literacy, Territoriality and Colonization* (Ann Arbor: Michigan University Press, 1995).

———. 'Espacios geográficos y localizaciones epistemológicas: la ratio entre localización geográfica y la subalternización de conocimientos', en *Pensar* [Bogotá], available at www.javierana.edu.co/pensar/Rev34, 1996.

———. *Local Histories/Global Designs. Coloniality, Subaltern Knowledge and Border Thinking* (Princeton: Princeton University Press, 2000).

———. *Historias locales/diseños globales. Colonialidad, conocimientos subalternos y pensamiento fronterizo* (Madrid: Akal, 2003).

———. *The Idea of Latin America* (London: Blackwell, 2005).

Milner, Jean-Claude. *L'Oeuvre claire: Lacan, la science et la philosophie* (Paris: Seuil, 1995).
Montgomery, Scott L. *Science in Translation. Movements of Knowledge Through Cultures and Time* (Chicago: Chicago University Press, 2000).
Piaget, Jean. *L'épistémologie génétique* (Paris: PUF, 1970). [1970a]
———. *Épistémologie des sciences de l'homme* (Paris: Gallimard, 1970). [1970b]
Plata, Sergio. *Visions of Applied Mathematics. Strategy and Knowledge* (Bern: Peter Lang, 2007).
Pratt, Mary Louise. *Imperial Eyes. Travel Writing and Transculturation* (Abingdon: Routledge, 1992).
Ricoeur, Paul. *Temps et récit*, III (Paris: Seuil, 1985).
Sacco, Joe. *Palestine* (Seattle: Fantagraphics Books, 2001).
———. *Safe Area Gorazde. The War in Eastern Bosnia 1992–95* (Seattle: Fantagraphics Books, 2005).
Sarlo, Beatriz. *La imaginación técnica: sueños modernos de la cultura argentina* (Buenos Aires: Nueva Visión, 1992).
Sebald, W. G. *The Emigrants* (London: The Harvill Press, 1996).
———. *Vertigo* (London: The Harvill Press, 1999).
———. *On the Natural History of Destruction* (London: Hamish Hamilton, 2003).
Solms, W. 'The Problem of monomania. I. Poriomania', in *Wien Z Nervenheilkd Grenzgeb.*, 11(1): 43–77, 1955.
Soundaramourty, Radjou. 'De l'errance, comme expression pulsionelle soutenue par un fantasma', in *Analyse freudienne presse*, 6/2003, 90–100.
Spitta, Silvia. *Between Two Waters. Narratives of Transculturation in Latin America* (College Station: Texas A&M University Press, 2006).
Tönnies, Ferdinand. *Community and Association* (London: Routledge and Kegan Paul, 1955).
Vera Tudela, Elisa. 'Review on *The Darker Side of the Renaissance*', in *The Journal of Modern History*, vol. 69., num. 4, 1997, pp. 863–864.
Verón, Eliseo. *La semiosis social* (Barcelona: Gedisa, 1987).
Virilio, Paul. *Esthétique de la disparition* (Paris: Galilée, 1989).
———. *Guerre and cinema. Logistique de la perception* (Paris: Cahiers du cinéma, 1984).
———. *La machine de vision* (Paris: Galilée, 1988).
———. *L'art du moteur* (Paris: Galilée, 1993).
———. *Un paysage des événement* (Paris: Galilée, 1996).
———. 'Interview with James Dear Derian', in Speed 1.4, http://proxy.arts.uci.edu/~nideffer/_SPEED_/1.4/, 1997.
———. *Unknown Quantity* (London: Thames and Hudson/Fondation Cartier, 2002). [2002a]
———. *Ce qui arrive* (Paris: Galilée, 2002). [2002b]
Wallerstein, Immanuel. *Historical capitalism with Capitalist Civilization* (London: Verso, 1995).

———. *European Universalism. The Rhetoric of Power* (New York: The new Press, 2006).
Weber, Max. *Wirtschaft und Gesellschaft* (Tübingen: Mohr, 1925).
Wright, Georg von. *Explanation and Understanding* (Ithaca: Cormell University Press, 1971).
Zumthor, Paul. *La Mesure du Monde. Représentations de l'espace au Moyen Agê* (Paris: Seuil, 1993).

Chapter 5
Reversal Thinking

> The idea of isolated systems in science was abandoned long ago. That is also why I believe the notion of 'autopoiesis' [from Varela and Maturana] is an interesting one. Equally, in the domain of culture and thinking in Latin America we should abandon the time-consuming task of trying to create the perfect isolated system. Nothing as such exists or, if it could exist, it will require such amount of sources, that will be inaccessible anyway for local authors and thinkers. We should do what is in hand for us: corrupt, distort, change, modify the Western/European tradition of thinking not because it is a way to oppose to it – because it represent, for example, the colonialism – but because is a way to construct a local domain. Instead to slow down and question the past, we should speed up and look to what practically can be done towards the future. Culture is not a literary art, not even a literacy-orientated activity anymore. It is an activity that product objects and things, therefore what matters is not what they mean but what they can do, how they can operate. I believe this is the profound connection between our present and the pre-Columbian civilization. And I see it not as an intellectual opportunity or as a new area for research but as a practical possibility to put to work many contradictory present elements and a number of local aspects of the local environment.
>
> Elias Ingaramo (1992: 122)

In the previous chapters we have explored the issues of thinking, culture, understanding, and knowledge within the Latin American domain from various viewpoints but mainly on the assumption, as indicated at the beginning, that Latin America is a concept or, more precisely, a particular type of conceptualization that acquires the form of space. This led us on to an exploration of many conceptual aspects and this is the moment to briefly evaluate the possibilities and/or consequences that such directions and aspects might imply and/or generate, not only regarding the definition of 'Latin America' itself but also with regard to how that notion should/could be located.

Three further aspects demand then a brief development before we can conclude. Firstly, an exploration of the connection between what we refer to, following previous works (see the classic linguistic work of Benveniste, 1966; also the always useful Bourdieu, 1982; and the more specific work of Mignolo, 2000a and 2000b), as the locus of enunciation and the idea of writing. Secondly, a commentary that revisits what was already said about thinking as determined by a spatial construction. And, finally, some concluding remarks about the notion of 'reversal thinking' that has been discussed throughout this work.

1. 'Writing from the Colonies'

A profound study of the pre-Columbian and pre-Western/European domination of Latin American in terms of ideas and culture is indeed fundamental. The work that has been done already, even when interesting and groundbreaking, is certainly not enough (see for example Brothersthon, 1979 and 1992). However, from the local and present point of view, this potential study or studies will have the secondary role of considering the colonial perspective that has dominated the notion of knowledge in Latin America for the past four centuries. Because it was always considered to an appendix or peripheral part of Europe, 'America' did not enjoy the benefits of the *alterity* ('otredad', otherness) of other colonial regions such as Africa or Asia. Even worse, since the seventeenth century 'Latin America' has also been an appendix of 'North America'. And this *double marginalization* has also become a mark of distinction of all local cultural developments. As we discussed in Chapter 3, in this sense colonialism is less an ideology or a political idea than a way to be located (individuals, objects, institutions) within a particular spatial design or, more precisely, within a particular form of establishing the spatial meaning of what is understood as local. Therefore, a critical and radical perspective of Latin America's cultural evolution includes a regular consideration of this epistemological situation by which concepts, ideas about knowledge and learning, as well as several definitions of local culture, are determined by formulations about space.

'Writing from the colonies' means writing from a particular geography and from a particular environment and it is not appropriate in our view to associate this situation or condition with a particular topographical map, country or nation. This is the great transformation within the post-territorial condition (see Canaparo, 2005: 251 ff.): the meaning of 'colonies' is something that is *also* mobile. Of course, in many ways it remains attached to a particular physical space but the meaning of that space is sometimes not within the space itself. Furthermore, with all the regular movement of people (immigration, emigration, travelling) that has become a way of life in our time, the sense of 'colonies' has also been radically transformed. The fact that the meaning and conditions of knowledge and understanding of what is considered as local is determined by something that happens outside that local space is not only a result of the colonialism discussed in Chapter 3, but also of the movement of people, of the migration element that we discussed in Chapter 2.

And if we concentrate on 'writing' it is not only because of the central role that writing had (i) in the pre-Columbian period (see for example Brotherston, 1995), (ii) within the conquest and (iii) still has in Latin America, but also because (a) writing has been transformed into a less letter-orientated activity (and therefore into getting closer to a sort of pre-Columbian condition, at least in the historiographical sense), and (b) because the notion of authorship has been developed in the Latin American domain to such an extent that it has become a sort of public paradigm of subjectivity. Or, more specifically, what for Western/Europeans is represented by the notion of the individual in the classic Freudian sense, is in Latin America often established less through an 'inner perspective' and more through a public and commercial view of what an individual is, which fits with the local development of the concept of authorship.[1]

1 For a discussion of this, focused mainly on the literary sense and editorial market see Canaparo, 2000. How the construction of authorship becomes a public and social paradigm in the constitution of subjectivity is something that is profoundly entrenched in the local definition of culture itself. And it is interesting how, following Western/European authors like Foucault's idea of authorship and knowledge, local academics can go in such opposite direction (see Foucault, 1969 and 1971; and also Abraham, 2002): it is not in relation to academic/scientific knowledge in the Western/European tradition, as Foucault suggest for Europe, that subjectivities are formed locally. A good analysis of this condition and situation has yet to be made.

'Writing from the colonies' is not a political stand or an ideological perspective, but a position about knowledge. And less than a position about colonialism itself, it is an attempt to make the 'colonial difference' – as W. Mignolo describes the effect of colonialism within the colonies (see for example Mignolo, 2000a and 2005) – something from which things can be produced, even when criticized. 'Writing from the colonies' is then not a choice but an imposition, like the family relatives we have or the topography of where we were born, something that cannot be erased not only because it is part of what we are as individuals but also because it is there not for today or yesterday, but is a constitutive part of the present notion of culture, time and knowledge. To construct a radical thinking even when we are conscious of this starting point is a crucial part of what we call 'reversal thinking'.

1.1 Understanding Writing

Although many authors have dealt with this topic, the works of W. Mignolo are perhaps the most complete and accurate up-to-date conceptual study of the notion of writing during the colonial and pre-colonial periods (see for example Mignolo 2000a). His persistent arguments about the local conditions of knowledge and writing in relation to those imposed by the conquest are indeed a landmark in the spheres of historiographic and cultural analysis. However, as indicated before, this interesting analysis has no direct functionality or implication in the constitution of the present environment of what we call Latin America (see Chapter 2), and even less in what we classify as 'culture' (see Chapter 4). Of course, the reason for this is not merely that there is a colonialist perspective still in place locally, as some authors might like to suggest (see for example the well known Galeano, 1973).

The writing and knowledge of local conditions as described by Mignolo regarding pre-Columbian communities have gone forever due to the conquest, colonialism and, later, because of the cultural and financial dominance of the Western/European perspective/paradigm. What remains possible and interesting is not so much a return or 'revalorización' (re-valuing) of this *impossible past*, but rather an exploration of the consequences, however minor, that these original local conditions and situations produced in the main Western/European perspective/paradigm as implemented locally. This exploration and interest does not necessarily specifically identify a number of characteristics or aspects but sees where this *twisting and degradation* of Western/European traditions leads the locals. In fact, the conception of the present as a form of *degradation and corruption* is probably the clearest result of the entanglement between original local conditions and colonial perspectives imposed since the fifteenth century.[2]

If it is possible, in the historiographical sense, to distinguish 'two sides of the letter', to use Mignolo's metaphor (see Mignolo, 1995), to refer to the clash between local tradition and Western/European traditions from the fourteenth to the seventeenth centuries regarding the activity of writing,[3] it is not possible to make the same distinction within the present or within the immediate environment. It is paradoxical that this historiographical approach, for reasons that Mignolo himself has developed quite well, can only facilitate an understanding of the present when the model assumed is based on the Western/European sense of 'logocentrisme' and literacy. The impact of historiographical analyses like the ones carried out by Mignolo on his comments can only be considered within a community for which literacy and alphabetic writing is still an important component of the definition of culture itself.

The analysis becomes even more complex when we start to consider that the Western/European tradition itself is moving towards a less letter-orientated notion of writing (see for example the classic works of Derrida, 1967 and 1978; and Virilio, 1984, 1988, and 1993). Therefore, it is not evident how useful it would be to attempt to base the analysis of Latin America as a concept or conceptual evolution on a potential re-evaluation of the pre-Columbian conditions of knowledge as defined by a historiographical work. It is in order to avoid any confusion in this sense that we believe it is important to establish a distinction between

2 On discussing other issues but in a similar context, Alejandro Kaufman maintains: '…lo que busca el proyecto de la escritura es cómo escribir una frase que se sostenga en estado de discussión, o de resistencia, o de interpelación, inclusive de supervivencia…' (Kaufman, 2007: 106).

3 According to Mignolo's viewpoint, this 'two sides' includes/refers also the subsequent local evolution following the 'clash'.

what is worth doing at the historiographical level and what can be done at an operative level in terms of defining an environment and a local sense of the present.

1.2 As a Historiographical Question

Of course it is possible to implement Mignolo's suggestion and proposal, about Latin America as a concept and cultural phenomenon, by questioning the Western/European tradition of knowledge (see for example Mignolo, 2000a and 2005) and somehow rescuing the obliterated/hidden local perspectives in cultural terms. However, this is clearly a historiographical issue. It is, as we have discussed already (see Chapters 2 and 3), a way to question a particular notion of knowledge, learning, literacy and writing, as developed in relation to Latin America, but it is not a direct method for forming the present understood as environment (see Chapter 4).

Writing from the colonies, as a historiographical question, allows us to generate a re-orientation in the way we can conceptualize locally the Western/European tradition that has dominated the notion and idea of Latin America since the sixteenth century. In this sense 'historiography' is a crucial tool when discussed together with the idea of writing, yet it does not have a direct impact on the way the notion of present and the environment are constructed locally. It is only when historiography becomes a *cultural object* and, therefore, is able to generate different meanings and images, that it becomes useful in the analysis of the local environment. In fact, it is due to this possibility that Mignolo's works have appeared in the last ten years. However, a further paradox has emerged, since it is only in literacy-orientated cultures that historiography can gain such status.

As Hayden White (1928–) and other authors have already argued, the fact of considering historiographical works as relevant and crucial in understanding and changing the notion of culture of a particular community is at the centre of the Western/European tradition (see for example White, 1978 and 1987; McKenzie, 1986; and Petrucci, 1999). And it is in this sense that Mignolo's work is not only paradoxical but also situates him in a local tradition of authors who try to deal with a dissociate past that denies the Western/European tradition but takes it as a source for conceptualization. What is paradoxical is the fact that Mignolo's work is inscribed in a Western/European tradition for which historiographical analyses are considered to have an impact on the notion of culture that a particular community has about itself, but within his arguments Mignolo avoids and ignores the relevance of this condition when defining what Latin American as a concept or conceptual evolution might be.

1.3 As a Location Point

Writing from the colonies, as a location point or, more clearly, as defining a *locus of enunciation*, also goes back to a clear Western/European tradition of understanding understanding. However, Mignolo's point – which we believe to be pertinent – is not about who conceived the idea, but about the fact that this situation should have become an epistemological priority when defining what Latin America is.

Writing from the colonies, as a location point, allows us to develop the location of knowledge as mentioned at the beginning of this work (see Chapters 1 and 2). Furthermore, on working around this location, 'writing' acquires a new meaning as the traditional Western/European sense of writing has been modified and transformed (see for example Mignolo, 1994a and 1994b). Writing in this context is more closely connected with visual practices than with textual domains.

Furthermore, what 'local' means here is not necessarily connected with a particular topographical formation. The 'local' is a construction that, in principle, can happen anywhere, but, because of the conditions that it requires, it clearly does not happen everywhere but just in specific constructed places. What should be clear here is the fact that what we consider 'local' has many forms and nowadays is no longer composed of just one single place, contrary to what has traditionally been suggested at the local level by the same authors that postulated modernity and industrialization as the aim of the local States (see for example Martínez Estrada, 1969; and Prebisch, 1963). The distinction we made in Table 2.1 (Chapter 2) between space as land and space as homeland, as well as the forms of imperial space described in Figure 3.4

(Chapter 3), and the four forms of space indicated in Table 4.1 (Chapter 4), all point to the same thing: the construction that today we consider as our 'local space' is not founded on any ideology, political concept or notion of culture as defined by literature and/or literacy.

Writing from the colonies in this sense means understanding writing – in terms of 'graphism', 'visualism', 'cinematism' – as standing in space. But it also means understanding how, by degradation, consumption and corruption, the local 'graphic' idea of writing and the letter-orientated notion of Western/European writing have been melted down and have ended up within a paradigm of planetary scale where 'culture' – understood as immediate environment and even as 'life' – is dominated by visual elements. That it is a complete waste of time to attempt to retrospectively modify what has been done historically, rests on this situation of *natural* confusion and degradation. The radicalism of Latin American thinking – if such a thing could ever be possible – relies more on a frantic race toward what can be done than on a hypothetical historical settle/settlement.[4]

1.4 Writing as Standing in Space

Writing, then, is not exactly writing in manual terms but understood as a more general graphic and visual activity. As we have already suggested elsewhere, the classical understanding of writing as a literary-orientated activity ('las letras') has more and more become a cinematic activity or, more precisely, a cinematographic-orientated activity (see for example Canaparo, 2005: 121–150 and 205–219).

To understand how the local 'graphic' orientation of writing and the letter-orientated notion of Western/European writing has ended up, by degradation and corruption, at the present notion of writing – that is clearly 'visually'-orientated – is probably a task for the next generation

4 And this is where we see one of the biggest limitations of the proposal with regard to the concept of Latin America, like the one coming from the perspective of a 'theology of liberation' or the one about 'border thinking', originating from authors like W. Mignolo (see for example Mignolo, 2000a).

of local academics. However, what is more immediate is the fact that any attempt to retrospectively modify what has been done should go into the category of 'time wasting'.

It is clear now that the idea of 'colony/colonies' does not have an ideological or political definition for us but we consider it as a form of location within the space. Less important than the State that drives the colonialism is, in our view, the idea of knowledge that not only allows the development of colonialism but that also makes the legitimate existence of such a State possible. Colony/colonies are precedents of colonialism: the idea of colonialism only exists in relation to a particular perspective about space. Furthermore, with the recent financial forms of colonialism this need for a particular form of space also became *imaginary*. This book is an attempt to briefly approach and discuss these spatial stands. The sense of the constructed and/or the manufactured that one finds nowadays in the colony/colonies is indeed rooted in this condition of existence.

2. Space of Thinking

Previously we have clearly referred to a 'space of thinking', 'to think spatially' or 'thinking space' as an area were the local is developed, not as a definite alternative to Western/European culture but as a *degenerate product* of it, a hybrid result of many aspects and elements (see mainly Chapters 1 and 2). More importantly, we have indicated that, even if we wanted, it is not possible in the present global conditions, to aim at or achieve any autonomous status in terms of languages or social organization. Such a situation – which was the aim and purpose of many *causes* and ideologies in the recent local past – is no longer available. However, these limitations do not prevent us from obtaining a clear idea and local perspective of the cultural and financial dependencies and the current political domination.

The space of thinking as understood by us is a domain where *the local* is produced and, no less important, where the local produces objects, artefacts, mechanisms and ideas. Because, whatever colonialism or kind of dominance one might consider, there always exists something that we call *local*. And much of this book was aimed at approaching that sense of *local* and, indeed, all the derivative elements that it generates, such as locality, location, localization, etc. From this perspective the assumption is that the more efficient these objects, artefacts, mechanisms, and ideas are within the immediate environment, the more efficacy ('eficacidad') and 'long life' ('larga duración') they will get outside that immediacy/environment (international, global market). This aspect, as we said before, is made relevant by the fact that, on many occasions, the local efficacy ('eficacidad') of some objects, ideas and artefacts is based principally and paradoxically on their circulation outside of what is considered local.

We have already discussed the distinction between philosophy and thinking as a way of differentiating between the scientific, professional and formalized form by which Western/European knowledge acquires legitimacy and the twisted, hybrid and foreign-orientated perspective by which what is locally considered as 'philosophy' – better indicated as 'thinking' – can be interpreted and/or understood. It is now time to go a bit further down this road – by juxtaposing this notion of thinking (Chapter 1) with the idea of space (Chapter 2) and the perspective of culture (Chapter 4) – and briefly observe what emerges.

'Thinking spatially' was described as an activity that occupies the space at the same time as it constructs it (see Chapter 1). It was also considered an activity where the borders and limits are constantly being negotiated. Furthermore, the entanglement between a sense of ground and the perspective of thinking spatially is established precisely from the fact that conceiving an environment and defining such an environment are one and the same thing (see Table 1.2). The local transformation/degeneration of philosophy into local thinking is also a crucial epistemic point in the location of knowledge. More precisely, the impossibility of producing 'philosophy' within a peripheral context is one of the characteristics of the sense of post-territoriality that permeates the whole local present. Thus, 'the space of thinking' is a combination of three elements: (i) a domain regularly established in a borderline environment, (ii) a clear relevance of a visual perspective, and (iii) a construction of an *imaginatio* as a 'world' and organization of knowledge (see Table 1.4).

We have also examined why and how the notion of space, entangled with the principle of thinking as an epistemic approach, results in a sense of culture that defines all its elements in a 'atmosphèrique' way. And this is relevant because it discards all possibility of making historical or time-scale interpretations of the culture and/or its intellectual evolution.

In fact, for an idea of culture founded on the notion that physical systems and reality are synonymous, and on the assumption that space is defined by the establishment of a geographic domain, in such a classical perspective (see for example Canaparo, 2005), it is inevitable that the very foundation of the local State and the principle of colonialism are part of the same historical and argumentative domain (see Figure 2.1). The fact that many educators featured among the politicians that created these States is certainly proof of it – as equally is the initial concern of these States with alphabetization, scholarization and education.

And if the local thinking and the formation of the State are connected it is also because, where an idea of knowledge is concerned it is the situation described above that emerges. The creation of the local State was generated by 2 types of colonialism ('double colonialism'), one of which came from the Western/European powers and the other one from the local authorities themselves. And so, in this context it is that 'space of thinking' or 'thinking space' that is considered like the space in its pure manifestation (see Figure 2.5). Authors like Ezequiel Martínez Estrada, who associate culture, knowledge and space, are clear examples of this (see for example Martínez Estrada, 1948 and 1957). And this is how thinking and environment have become naturally associated in local terms.

2.1 The Idea of Latin America

In 2005 Walter Mignolo, an Argentinean-born north American academic, published what I believe is the most relevant discussion about the notion of Latin America since the works of Ezequiel Martínez Estrada (1948) and Angel Rama (1926–1983) (1982). It was first published in London, in English, under the title of *The Idea of Latin America* by Blackwell (see Mignolo, 2005). The virtues of Mignolo's works are very clear and strongly developed: a clear theoretical approach, an innovative perspective, a real multidisciplinary standing, a commendable effort to revisit the local intellectual history of Latin America, and, perhaps most relevant of all, the proposal that cultural issues are also epistemic and conceptual problems.[5]

However, there are two or three elements in Mignolo's works that attract our attention. First, the fact that the relevant term of *locus of enunciation* should refer/ be applicable not only historiographically – as is the case in Mignolo's book (2005) – but also with regard to the present in the sense of 'culture' or 'marché symbolique', as indicated for example by contemporary Western/European sociologists like Pierre Bourdieu (see for example Bourdieu, 1992). What Mignolo fails to address more deeply is exactly this *present* dimension of the location of knowledge. If the authorship position/standing is relevant for the purpose of analysing the sixteenth century and nineteenth centuries, as Mignolo showed, why should it not be relevant to Mignolo's own authorship position? This is where we proposed following a more constructivist approach, which seems to be a bit more radical and coherent in this respect (see Chapters 1 and 4). As Mignolo himself sometimes seems to suggest, we must not only ask the authors how they interpret their work, we also need to explore how they *circulate* in the market. The case of Gloria Anzaldúa's (1942–2004) *Borderlands/ La Frontera* is, I believe, a paradigmatic one (see Mignolo, 2000a and 2005). Mignolo's analysis and interpretation follows Anzaldúa's claims, expressed within her book and elsewhere, but fails to indicate, despite the declared epistemic purpose, how her book *circulates* within the market, and, more importantly, fails to explain how and why this circulation is relevant and, in some cases, determines the ultimate meaning of the book. Anzaldúa's book circulates mainly in the US and almost exclusively within the university and academic communities.[6]

5 Of course, this idea is not new and had been originally developed by authors like J. Piaget (1937 and 1970) and E. von Glasersfeld (1995), among others. However, Mignolo was the first author to fully develop it in relation to Latin America.

6 To give an example, I can say that it took me more than six months, in London, to obtain a copy of Anzaldua's book, using the biggest internet book provider in the USA.

In many cases Anzaldúa's book is present in academic syllabuses not because of her ideas of knowledge or language but primarily because she declared herself a lesbian and feminist. The North American academic's interpretation – with the obvious exception of Mignolo – tends in that direction (see for example Keating, 1996 and 2005). Of course, there should be no problem with this largely extended bio/autobiographical approach; what I see as problematic is that this *reception* of Anzaldúa's book did not feature in Mignolo's analysis. And, of course, Anzaldúa's book is only a single example. The same could be said about many of the other cases that Mignolo brings/uses to illustrate 'the present'.

Secondly, following on from the previous point, it seems evident that we cannot criticize Western/European authors for being Western/European-orientated in their approaches and ideas. We can criticize them, when considering a foreign subject, for not been conscious of the epistemic limitations of their approach. But we cannot assume that their 'philosophical thinking' (in Western/European terms) should be social-orientated: philosophers are not religiously committed people and, more precisely, are certainly not ethical individuals by default. This can be seen with many academics nowadays, who are developing their lives (one pattern, one set) and following a completely different professional route (another pattern, another set).[7] The fact, for example, that many academics who have specialized in Latin America as a *subject* work within colonial universities and institutions in the USA, does not seem to generate or produce any conceptual or theoretical consequences to them.[8]

Thirdly, the excess of repetition that can be found in the diagnosis in Mignolo's work does not escape our attention (perhaps more evident in Mignolo, 2005). And it is not by chance that this excessive emphasis

7 And this is also where I see Mignolo's work as *too academic* because it implicitly assumes a close connection between biography/autobiography and authorship, that we consider to be much more problematic and less evident. In relation to this argument see Canaparo, 2000.

8 Of course, as indicated before, we are not referring here to any ideology or biographical issue but more to a practical and functional aspect: every author thinks and writes *from somewhere* and it is this 'from somewhere' which, in our view, cannot be avoided.

on diagnosis within a particular analysis has already been identified by P. Rovatti and G. Vattimo, in their classic work *Il pensiero debole* (1983), as one of the current characteristics of Western/European culture and thinking. It is interesting to observe that a book whose contents confront the Western/European tradition of knowledge, paradoxically endorses this heritage by its paratexts (location/placement of the book) and academic style of writing.

Finally and more generally, it is difficult to see how Mignolo's comments on Enrique Dussel (1934–), Rodolfo Kusch (1922–1979) and other non-Western/European theoretical authors, from the perspective of 'liberación' (liberation), are compatible with his theories about the location of knowledge, the hypothetical relevance of pre-colonial cultures and their conceptual implications. The 'contenido indígena' of the arguments deployed by Kusch and others are at no stage generated *from* a new idea or notion of writing and knowledge as Mignolo claims or, to be more precise, as he is certainly and paradoxically doing on his work. Kusch and others are still discussing historical perspectives and, antagonistically, Western/European conceptual systems (see for example Kusch, 1977). There is an *epistemic gap* between Mignolo's proposal about knowledge and the rather argumentative dimension of the works of *teoría de la liberación* ('liberation theory') – perhaps with the isolated exception of some of Dussel's writings (see Dussel, 1985) – and those of other authors who have been trying to establish a local autonomous cultural and theoretical point of view (see for example Kaufman, 2007). The fact that Mignolo somehow tries to place/inscribe his work within this tradition of 'local thinkers' makes the situation even more evident (see Mignolo, 2000a).

Therefore, in an attempt to avoid this problem and to implement a practical approach, we consider that a concept is a non-personal instrument, an independent artefact, a constructed device, as Bruno Latour for example has already shown in Western/European terms (see Latour, 1993). The meaning of a concept does not come from the authors that devised/exposed it, even when the author's biographical account could add many connotations to it. The historiographic entity of some authors should not be confused with the viability of some concepts (see for example Bourdieu, 1984). And, just as importantly, as we have already stated (see Chapter 4), a concept should always include its own history, situation and evolution if we want to have an accurate culture location and epistemic approach (see Canaparo, 2005: 165 ff.).[9]

The fact that what we call 'Latin America' is increasingly something manufactured *outside* the traditional topographical boundaries of what makes the land of Latin America itself is not new and, as Mignolo quite rightly indicates (see Mignolo, 1995 and 2000) and as we have discussed already (see Chapter 1 and Chapter 2), it cannot be dissociated from the way in which Latin America was 'discovered' by Western/ European people and settled as a colony.

Is not by discussing/exploring the evolution and development of the idea of Latin America that we will obtain some creativity and freedom from the epistemic and conceptual point of view. It is necessary to go beyond and, in first instance, see how *the idea of the idea* of Latin America works. And this is also what explains the previous discussions about 'constructivism' (see Chapter 4) and, in a minor way, the perspective of cybernetics (see Chapters 1 and 2).

2.2 The Idea of the Idea of Latin America

What does 'the idea of the idea of Latin America' mean? It indicates that what is relevant in relation to the use of 'Latin America' is, above all, a conceptual employment. The important aspect of study is not so much a historical account of the idea of Latin America but rather how Latin America became *an idea*. The idea of the idea refers also to a form of analysis where the epistemic aspects are considered as more relevant

9 An author can be considered as an 'actor' in the Latourian sense (see Latour, 2005), that is, as an artificial entity that does not respond to the biological person that is normally attached to it (see also Canaparo, 2000). But also as an entity that acquires meaning and physical functionality by its relation with a system or its network (see Latour, 2005: 63 ff.). And this, among other reasons, is why the ideological-orientated perspective is not appropriate nowadays: because it used to assume that any *authoritative* control was *absolute*. Our criticism of Mignolo's very interesting work also stems from this view. We cannot be *innocent* with regard to this: the consequences of an author's strategy cannot be measured by the views of a biological person (see also Canaparo, 2000).

than other folkloric-orientated elements. And, as part of this approach, the 'lugar de enunciación' (locus of enunciation) of the author-analysis becomes part of the analysis itself.

This in part is the natural realm of Mignolo's writings and in that sense his work on Dussel and others is very interesting and useful, as since it attempts to construct an intellectual evolution towards this idea of Latin America, which is the first one, at least in cultural terms, since the works of Martínez Estrada and Angel Rama (see Mignolo, 1995, 2000 and 2005).

And this is also why we maintain that the idea of Latin America, more than ever before, is actually connected to academies and corporations such as the universities. Therefore we are no longer dealing with some anthropological sense of culture that produces a meaning of what the idea of Latin America is, but, quite the contrary, that Latin America has become a manufactured perspective that varies and undergoes modifications according to its relation with the media and the production of universities. Thus, in a similar way to what used to happen to some cultural-State orientated Western/European institutions during the nineteenth century, the shift made in North American universities, from colonialism to imperialism, especially within the second half of the twentieth century, has become the new instrument of the cultural and historiographical domination of peripheral areas like Latin America.

For this reason we can only regard suspiciously the views of some academic authors – based mainly in North American universities and institutions – who are trying to transform issues like the so-called 'latino' subject/movement into a driving element of the whole idea of Latin America and of what is understood as Latin American space (see for example Grosfoguel/Maldonado-Torres/Saldívar, 2005). Is this academic attempt another form of colonialism/imperialism concerning Latin America? Is this 'latino' dimension the future of Latin America as an idea? It is perhaps too early to tell but what is clear for now is that, outside the US academic environment and outside the US territory, the 'latino' issue is not as interesting and relevant as some authors would like us to believe (see as before Mignolo, 2005: 95–148). The commercial success of some films, music clips and general literary books inscribed/registered with the 'rótulo latino' (latino label) is not enough to confirm this 'new internal colonialism' or promising future. The strong media phenomenon that the 'latino' issue has become inside/outside the US environment, is still not enough to replace the sense of culture upon which the idea of Latin America has been built in the last 400 years in conflict with Western/European traditions. And this is true, even when some academic and commercial spheres, like literature and music, have been completely over-written by the activities and views of the US market, universities and academics.

2.3 The Cinematographic Core

The consideration of cinematography as a natural and current form of reflecting (thinking) about our present, environment and historiographies does not result in a de-legitimization of the activities of writing as discussed here. In fact, quite the opposite, cinematography supposes such activity as a *realm*. And this is where we take stock of the many works done on this subject as our understanding is that the classical perspective of writing ('la ciudad letrada', 'las letras': Sarmiento, Bello) is not actually hostile or opposed to the graphic, visual and cinematographic views we are presenting, but quite the contrary, they work together and it is difficult to think – at least within the Latin American context – of one without referring to the other. However, what is radically different is the fact that, as already mentioned, this writing/written perspective is interpreted from a social and cultural context that is visually orientated.

The fact that cinematographic perspectives and schemes have been studied as 'cinema' or, more recently, as 'film studies' does not prevent us from considering cinematography as a more general and relevant form of understanding. Henri Bergson was probably the first Western/European philosopher to consider the speculative consequences of the development of cinema – at least that is the clear proposal in the classic cinematographic works of Gilles Deleuze (see for example Deleuze, 1983 and 1985). Peripheral areas like Latin America (see Chapters 1 and 2) cannot avoid this general situation, but even more relevant in this regard, the cinematographic mechanism is at the core of the constitution of the modern State and communities in Latin America (see Canaparo, forthcoming: Chapters 7 and 8).

This consideration of cinematography as a natural form of thinking is also directly related to the epistemic and conceptual aspects of 'the idea of the idea' understood as current cultural phenomenon. Cinematography has the necessary elements to deal with it. But more relevant, it is because cinematography is better at dealing with the incommensurability between perception and knowledge that has been considered, in various ways, as the major philosophical problem of modern times or, more precisely, as the most accurate definition of philosophical modernity. Incommensurability between the present and understanding, facts and writing, the present and reality, phenomena and writing, to mention just a few of the equivalent notions of this terminology.

Cinematography forces the exhibition of objects, people and narrations in such a way that its *second order* dimension is always there in the form of meaning (see for example Foerster 1974 and 2003). The fact, for example, that the simple evolution of an idea ('Latin America') cannot simply be considered from the point of view of its content ('first order') but must also be considered with regard to how that idea became an idea ('second order'), is something that the semiotic condition of the cinematography's visuality covers all the time.

And this semiotic condition basically refers to the location of three semantic aspects: (i) a meaning situated at the level of the *content* of the image, (ii) a meaning situated at the level of the *form and collective* elements of that image, and finally, (iii) a meaning located at the axis of the previous meanings with a sense of (a) present, (b) continuity and (c) reality. We call this condition *the cinematographic core* of local knowledge.

And this privileged condition of cinematography in the establishment of knowledge explains why we attribute great relevance in Chapter 2, within the constitution of the local State, its attached community and the dominant notion of knowledge, to the role of cinematography (see Figures 2.1 to 2.4). More specifically, the really relevant aspect to consider is the emphasis/assertion in the visual dimension that the development of cinematography implies (see Figure 2.1). We have already explored elsewhere to what extent this visual dimension changes our sense of reality (see Canaparo, 2007 and forthcoming); the intention here is to briefly explore how this visual dimension works in relation to the formulation of the local space (see Figure 2.4). And this double effect on (i) the establishment of space and (ii) the establishment of the sense of reality is where the visual dimension predominantly functions or, more precisely, where we decide what is defined as local.

The cinematographic then, understood as a domain in the circumstances described here, is the finest expression of this space of thinking. And indeed it is the most radical form of questioning of the Western/European culture that has been dominant since the sixteenth century. This questioning did not necessarily go entirely as it was planned, but, on the contrary, appeared mainly as a natural and local evolution, modification, and transformation of the Western/European tradition by twisting, distorting and de-naturalizing this tradition itself.

In this sense cinematography implies that *we do not understand once but many times*. Cinematography as methodology is a constant re-acting process, a scheme that is continually undergoing revision and change. And this is why there is a connection between the thinking ('pensamiento') understood as a cinematographic stand and the notion of 'reversal perspective' proposed in this book. The sense of *ground* in term of knowledge cannot be sustained, in peripheral areas like Latin America, from an enlightened point of view. Therefore the pretension of a paradigmatic stance is an illusion that goes in line with the 'internal colonialism' that was predominant among the 'founding fathers' who pursued the local political independence from the Western/European powers. This explains the useless analysis of a Latin America understood in terms of modernity or modernization.

The cinematographic core, on the other hand, is not a system or a scheme that can be set/placed in a series of theoretical axioms or precepts. As we have discussed previously (see Chapter 4), following Humberto Maturana's epistemic propositions, we understand the cinematographic core as a *condition of possibility* – which is the reason why we have also discussed the relevance of the notion of *horizon d'attente* (see Chapter 4). Nor is the cinematographic core a reproduction of some recent Western/European philosophical perspectives (see for example Frampton, 2006), for the simple reason that such 'reproduction' cannot be achieved due to the financial and economic conditions of peripheral areas. But also because cinematography today is less an intellectual outcome of Western/European colonialism, but rather a product related to technology, market conditions and media situations, that is, a reality – or a sense of reality, to be more precise – present in any corner of the planet that can be assumed or understood as an existing space. And this is the profound root of cinematography in our present every day life: space can only be defined – and thereby *exist*, as we have argued in these pages – in relation to the cinematographic principles as explored here.

3. The 'Reversal Perspective'

The various meanings of the idea of 'reversal thinking' were discussed in Chapters 1 and 2 and, to a lesser extent, in Chapters 3 and 4. However, what we will now focus on briefly is the context for this proposed notion of 'reversal thinking'. One of the main problematic issues of recent local theories about Latin America, especially during the twentieth century, has been the fact that the majority of them based their *legitimacy* on (i) a political promise, (ii) a counterpoint to what is understood as Western/European culture/tradition, or (iii) the ideal and isolated condition of cultural production. The 'reversal perspective' tries to avoid these conditions of existence right from the beginning and, even more, it attempts to go in the opposite direction by embracing all the existing meanings and interpretations of the past without questioning their legitimacy but looking closely at their epistemic stand. The 'reversal perspective' is interested in the production of 'becoming' (understood as present), in the status and content of the *horizon d'attente*, but not in the polemics regarding the historiography's legitimacy or in the delimitation of the academic or editorial communities.

'Reversal thinking' then, as proposed here, is a way not to avoid the consequences of Western/European dominance but, on the contrary, to assume them and to start what we will like to call its open manipulation. And by manipulation we understand not only the fact that we *use* Western/European authors and/or ideas but also that we attach to them a different legitimacy or functionality to the one generated in their original context. This is the great *epistemic difference* that very few authors have been capable of recognizing or paying the attention that it deserves.[10] The question or problem is not whether we consider a Western/European heritage or not, because the possibility implied in such a question is denied right from the start since we cannot ignore that foundational/basic tradition and start from zero. A more pertinent question would be to question/modify the relationship that such a tradition has with its own context of production. And obviously another pertinent question would be to see how it works in relation to local space. We also call this the 'reversal perspective'.

The long cultural and intellectual 'inferiority complex' – one of the consequences of colonialism – that the local communities suffer is more related to their inability to do something with the cultural heritage of colonialism, than to finding some kind of miraculous *zero degree* of local culture and politics.

The location of knowledge is then, from this viewpoint, less a proof of potential analytical and informational capabilities and more a simple and immediate way of defining 'what is there' (question of localization), 'what its meaning is' (question of understanding) and 'what we are going to do with it' (question of knowledge). Thus 'reversal thinking' should be fully understood as a present form: it is not an attempt to go back in time or in cultural evolution but a pragmatic assumption of *what is there* (culture, historiography, tradition, foreignness) *and here* (territory, language, market, locality). We believe that the proposed 'culture of the accident' is not only an interpretation tool but also something that fits with the conditions of the pragmatic requirements of 'reversal thinking'.

3.1 Constructing Space

The proposed notion of *reversal thinking* cannot pretend to be original or new, but can claim to be innovative in the constructivist sense given to the word by Ernst von Glasersfeld, among others (see Glaserfeld, 1995). Something is innovative when a conceptual system, locally

10 From the *Comentarios reales* of the Inca Garcilaso to *Ficciones* from Borges, only a few authors were able to work at the edge of this epistemic difference.

based, modifies or challenges the dominant interpretations of the world-view about that local space (see for example Piaget, 1970). Less concerned about the status of ontology or the way in which reality is constructed (see for example Maturana, 1995a), 'reversal thinking' assumed that knowledge could only be constructed locally, which is one of the aspects – not always mentioned – of Glasersfeld's constructivism (see Glasersfeld, 1995). For Glasersfeld the innovation of 'constructing' does not come from the 'new' or that appears to be 'new' – an *illusion* after all – but comes from the modification which takes place (operates) on the existent. In this Glasersfeld follows Piaget's notion of knowledge (see Glasersfeld, 1995; and Piaget, 1970).

It should also be said that for the 'reversal perspective' – reversal thinking understood as an operation/process – we also assume that we do not understand once but many times, that is, following what we discussed in Chapters 2 and 4, that we not only make many interpretations of the same phenomenon but also that there are no theoretical or cultural conceptions that can stand for long periods and, therefore, that culture means constant adjustment. And, on the other hand, this situation fits with the sense of environment we discussed in Chapter 4.

How space is constructed was the subject of previous Chapters (see Chapters 2 and 3) and we will no return to it. The point here is the constructivist stand/condition of knowledge and thinking, that is, the fact that knowledge and thinking *cannot* be postulated outside the local environment or, more precisely, cannot be considered/discussed outside the atmosphere generated by the local environment (see Chapter 4).

3.2 The Travelling Theory

The 'reversal perspective' assumes that ideas and people are travelling all the time, therefore notions like exchange, translation and interpretation, have a relevant role in the constitution of the local space (see Chapters 1 and 3). However, what is substantial about this is the fact that on each occasion, on each trip or journey, at each exchange, a notion of what Latin America is is implied. In these circumstances the 'reversal perspective' appears as a reversal situation in the creation of a meaning for Latin America, as we try to illustrate in Figure 5.1. This is not a new approach and with minor differences it could be found implied in the 'epistemologías toponímicas' of the Inca Garcilaso for example, and likewise in the idea of 'liberation' as understood by Enrique Dussel (1985) and Rodolfo Kusch (1977).

The 'reversal perspective' in these terms is clearly visible in relation to the definition of space – as we have been arguing throughout this book – and to the establishment of an authorship – considering it, as previously stated, in relation to the central role that the notion of authorship has occupied in Latin American's cultural evolution. Authors like E. Dussel and R. Kusch – and indeed Mignolo – have a perfect understanding of this link with Western/European traditions in terms of authorship and in this regard their work should be followed and improved. Probably the clearest example of this is in relation to the editorial market. If one briefly analyses the course of publications of authors like Dussel or Kusch the authorship element is distinctly visible: they never publish in the established market place but always in a *oratio obliqua* condition to it and to the Western/European dominant paradigm.[11] More importantly, they maintain and/or suggest that there is a direct link between (i) the arguments deployed, (ii) the epistemic stand of these arguments and (iii) the way in which they are distributed in the market place or in a particular community.

11 There is another interesting case, worth mentioning because is it concerns an author that, unlike Dussel and Kutsch, and more like Mignolo, has made part of his career within the north American academia. We are referring to the Uruguayan cultural analyst Angel Rama. The two most reputed works of Angel Rama – *La transculturación narrativa de América Latina* (1982) and *La ciudad letrada* (1984) – which probably feature among the most commented on and quoted works of the second half of the twentieth century within Hispanic and Latin American Studies, were, however, published by two small publishing houses in Mexico and Hanover. Years after Rama's death, an extremely notorious North American academic publisher decided to publish one of these works in English (*The Lettered City*, 1996) and such an event became, for critics and analysts, a sort of intellectual consecration for the author. Of course Rama in English is not the same Rama published by Ediciones del Norte in 1984. This is not only because the text is not the same, but also because as cultural and intellectual objects the two books 'circulan de manera diferente'. And it is this point we are really addressing.

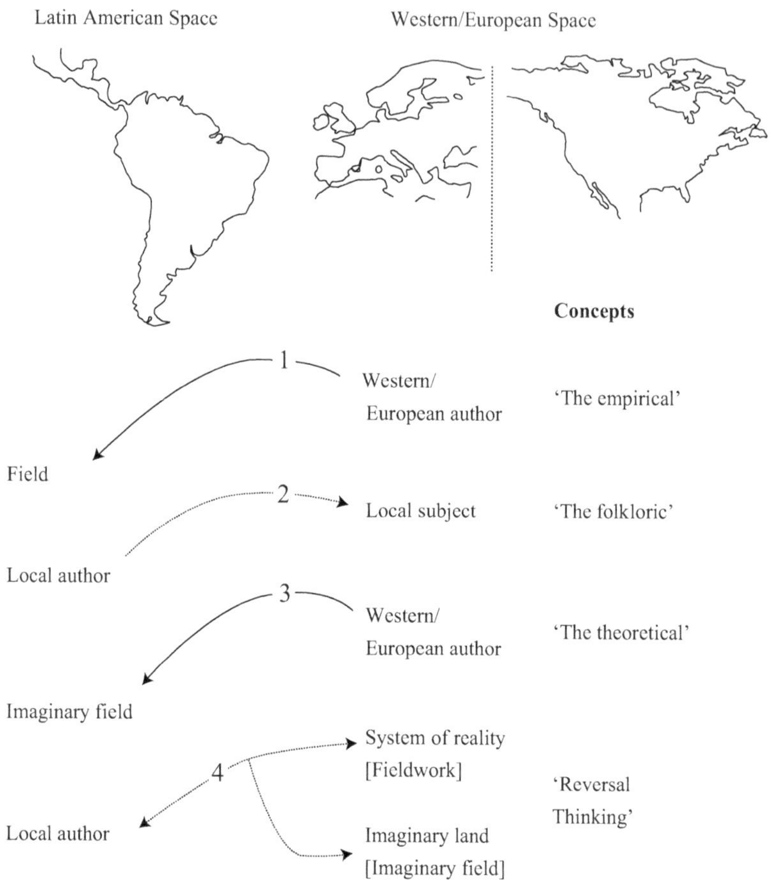

FIGURE 5.1. The 'reversal perspective' in terms of space and authorship.

As we discussed elsewhere in relation to the formation of space and local culture in Latin America (see Canaparo, forthcoming), travelling has a central role in the idea of Latin America itself. And it is also for that reason that this idea of travelling was initially where the 'reversal perspective' theory originated. And travelling cannot be dissociated from the evolution of Latin America itself as an idea and/or a concept. Figure 5.1 shows there are 4 trips/journeys that from this perspective create/develop the idea of Latin America based on the notion of travelling. And we refer to 'trips'/'journeys' as an ensemble that produces a travelling condition in the way we mentioned before – that is, as a driving element in a cultural evolution– ; in which case 'trips'/'journeys' refers not only to the authors who produced them but also to the constitution of the trip/journey as a separate entity, as an actor in the Latourian sense, as a concept (see for example Latour, 2005). And that is why these trips/journeys can be considered from two perspectives: (i) as historically orientated and (ii) as conceptually orientated.

The first relevant trip/journey, considered historically, was by the Western/European authors that 'created' and 'discovered' Latin America as Continent – from Amerigo Vespucci (1454–1512) to Alejandro Malaspina (1754–1810) and from Alexander von Humboldt (1769–1859) to Charles Darwin. These trips/journeys established the space and the space as a field, that is, the space that can be subject to study, analysis, description, taxonomy, etc. The second trip/journey, historically-orientated, is the one undertaken by local authors, within a Western/European context but dealing with a local subject as well – this refers to authors from Inca Garcilaso to D. F. Sarmiento and from J. Martí (1853–1895) to J. L. Borges. The third trip/journey, historically-orientated, that already assumes the existence of Latin America as an idea, is from Western/European authors towards an imaginary field. Authors from Jules Verne (1828–1905) to Italo Calvino (1923–1985) can be considered in this category. The fourth trip/journey, historically-orientated, has a dual purpose and it differs from the previous trips/journeys – which normally go in just *one direction* – by the fact that it goes in two directions, that is, it can refer equally to local authors and to Western/European authors. The dual purpose is towards a system of real (field work) and/or an imaginary land (imaginary field).

Understood from a conceptual point of view these trips/journeys also illustrate an intellectual and cultural evolution. The first trip/journey, viewed from this perspective, refers to/creates the *empirical*, that is, the sense that by exploring the world of Nature we create the space. The second trip/journey refers to the *folkloric*, that is, to a number of elements that, rather than constructing Nature give it a particular characteristic. The third trip/journey, that takes for granted the first and second trips/journey, refers to the theoretical, that is, the belief that only through a theory can we achieve an spatial understanding. The fourth trip/journey refers to what we call reversal thinking, that is, to the fact that all trips/journeys are *round trips/journeys*. More specifically, this trip/journey refers to the fact that, because there are no universal codes or languages, in the end every trip/journey goes to a particular location, which can therefore be understood as an imaginary place or also as a field of work.

Figure 5.1 offers a diagram of an historical as well as a conceptual evolution and, of course, the suggested order is only provisional and useful for the benefit of the explanation: today things are not as clear as they seem to be in these schematic proposals. However, the main elements are there and the trips/journeys, as an ensemble of the travelling element, are still functioning/acting. What is more evident today is, on one hand, the entanglement of elements and, on the other hand, the condition of *artefact* by which many elements of a cultural explanation work.

From the Western/European perspective the idea of Latin America has always been constructed/formed based on a narration – that also explains the pivotal role of authorship, as well as the central epistemic situation of the activity of writing. Whatever radical perspective local authors might have about colonialism, this situation/condition cannot be avoided.

Considering the forms of space discussed in Chapter 4 (see Table 4.1) and then following the evolution outlined in Figure 5.1, we can establish four different definitions of Latin America as a space:

First definition. Latin America appears as a 'field', as a place of research for Western/European authors. This is the first trip/journey where the ideas of world, newness, taxonomy, and exploration prevail – it is a trip/journey to an empirical domain. The form of space that prevails here is that of *place* ('lugar'), a basic form of ground.

Second definition. Latin America appears as 'local subject' – considered and used also by local authors – but established for Western/European authors and parameters. This is the second trip/journey where Latin America is constructed as a unity and obtains the status of reality. For the same reason this is where the folkloric dimension emerges as the purest expression of what is considered as local. As a complement of the previous definition, the space that prevails here is hat of a *place-world* ('ubicación'), which initially also acts as a basic form of ground.

Third definition. Latin American became an imaginary field for Western/European authors at the same time as the idea of 'autonomy' from Western/European traditions appeared locally. And this 'autonomy' from the traditional powers was first expressed in a theoretical form (political ideas, literary ideas, artistic ideas). The idea of *site* ('localidad', 'sitio') is the form of space that corresponds to this understanding of Latin America, which normally operates as a cultural ground.

Fourth definition. Local authors have not exhibited or postulated any notion of Latin America but have started to exercise/use a particular analytical perspective that implied the existence of Latin America as a concept and *épistéme*. Thus Western/European space becomes a 'field' for these local authors and later an imaginary field too. This is when the reversal trips/journeys start: Latin America is a trip/journey to Europe and also, no less significantly, a return trip/journey back. These movements can be repeated many times. Authors and situations have started to work under this 'reversal' mechanism: the 'reversal thinking' becomes a local methodology. In terms of space, this version of Latin America has two forms: one being '*milieu*' (surroundings, 'periferia inmediata', community) and the other being *environment* (atmosphere, weather, forecast), which can be situated respectively as cultural ground or epistemological ground.

Of course, these definitions of Latin America are not there to be associated with any particular chronological period or political circumstances. These are trips/journeys that are *melangé* and happen all the time in all directions. There is not one pure or isolated trip/journey that prevails. In this sense Latin America is indeed the result of an *oratio obliqua* exercise though not just from one element but from many. However, a sense of *evolution* always exists and whatever order they might appear in, all 4 of these trips/journeys are present in the constructed sense of Latin America. And this is what we had in mind when we proposed that the notion of degradation was the 'real' evolution in the thinking within Latin America (see above and also Chapter 2): the evolution of these trips/journeys constantly changes and corrupts what is travelling. Thus the 'reversal perspective' is the result of a series of degraded trips/journeys between the local environment and what is outside it (see Figure 5.2).

The meaning of *field* also needs to be adjusted as there is no past in the cultural-monumental Western/European sense (see Foucault, 1969) in Latin America. And that is why the Western/European sense of *research* does not have much relevance. What is considered as crucial and fundamental is something that was *artificially* constructed as a monument or, more commonly, is something that needs to be expressed within the *media* to acquire existence. The *public dimension* of things is the ontological borderline of the world for Latin American authors concerned with issues like culture, society, literature, etc. The travelling element is then tied to this original condition where the *monuments* of the community are generated in the public domain and not by any historical or evolutional process.[12]

Furthermore, it is possible to assert that from this general perspective the sense of travelling – which does not only refer to moving around or to translating – dominates the meaning of Latin America as space and this is also why we postulate an 'culture of the accident' as a local distinctive constitution. Thus travelling is a constitutional element of the culture or, more exactly, the way in which culture evolves. The notion of culture itself, associated with the idea of Latin America, was constructed from the very beginning as a travelling domain.

3.3 Migratory Trips/Journeys from Nowhere

There is then a direct connection between the notion of *migration* discussed in Chapter 2 and the idea of post-territory, mainly understood as a lack of any relation between the topological system and the semiotic/signal system deployed in the past to recognise it (see Canaparo 2005 and Canaparo, forthcoming). Furthermore, as has been discussed elsewhere

12 The common battle regarding to names on streets, squares and other public spaces of cities, to mention only one example, is typical of this condition.

(see Canaparo, 2005; and also Moraga, 1993), the very idea of *land* is at stake within the notion of post-territoriality. The ensemble of goods, technologies, objects, animals, architectural sets, and human beings, that composes the immigration/emigration condition, gets functionally entangled with the land in an *environmental* way (see Chapter 2). Therefore we are not alluding to some topographical configuration (mountains, trees, etc.) or administrative division (flag, local history, etc.). As Moraga put it:

> For immigrant and native alike, land is also the factories where we work, the water our children drink, and the housing project where we live. (Moraga, 1993: 173)

The *relativization* of the human being in the context of the environment, as with the Western/European tradition, could certainly be explained by the evolution of machinism and industrialization (Marcuse, Lebeau), from the humanist point of view (Camus, Mounier), and from an economic perspective (Marx, Hayek). However, we cannot avoid the fact that, as Maturana argued (1995a), our understanding, knowledge and conceptualization of what we do is entangled with the environment itself, understood as a general *milieu*. In this strict sense *we cannot think about the environment*: we live in it (see Maturana, 1995b). The 'anthropomorphical explanation' that places 'man' at the centre of any explanation (see Foucault, 1966) and that many authors still utilize in social sciences and cultural analysis does not fit with the present local situation of transculturation ('transculturización') and globalization (see, for example, Hart, 1999 and Fernández, 2006).

What the theory of travelling offers us is not the opportunity to construct some origins or roots in a Western/European sense, but a circuit, a series of movements through which the meaning of proximity (familiar, locality) and distance (foreignness, strangeness) is established and developed. And that is why the migration element, as discussed in Chapter 2, is relevant. What traditionally characterized the migration was the fact of departure from somewhere, the fact that someone or something was leaving a familiar entourage and going to some unknown *milieu*. This situation is nowadays one of the possible definitions of the *present* and not an unusual or exceptional circumstance anymore. If we interrogate our cultural activities in Western/European sense, rather than reproducing a colonial analytical model and argumentation, we see that we are going nowhere in spatial terms, when what we need is just the opposite: we need to travel, we need to create and develop places. Places acquire meaning less and less because they are considered *heimatland* ('país de nacimiento'): the sense of *pertenencia* (belonging) is almost completely constructed nowadays.

3.4 'Lo familiar y lo lejano'

And together with this notion of travelling and migratory condition, there is another constant and regular presence in the formation of an idea of Latin America: the question of what is considered as 'close' (familiar) and what is 'far away' (non-familiar). Strictly speaking we should talk about 'familiaridad y lejanía', that is, a sort of mechanism that both defines our space and its elements, defines what is strange, alien and distant from our space. However, this distinction is not exclusive as one aspect is connected to the other, as shown in Figure 5.2. And it is this *dependence* on the familiar and what is far away – and vice versa – that we call 'environment'.

The difficulties that many Latin American communities have with regard to the status of what is far away ('lejano', 'distancia', 'extranjero'), as we mentioned already (see Chapter 1 and Chapter 2), cannot be detached from this *constitutive element* of the local. The inability to define the meaning of strange ('extraño') or foreign ('extranjero') in a particular moment, as Alejandro Kaufman (1955–) (2007) has already pointed out, reinforces the relevance of the mechanism of a *horizon d'attente* (see Chapter 4) as the main domain in determining what is considered as *legitimate* local knowledge.[13]

13 'El rol de la educación en la Argentina en relación con el conocimiento, el proceso de socialización del conocimiento, imagina al conocimiento como algo que ocurre en otra parte y en otro momento. Y esto es estructural. Y nuestras vidas también están inscriptas por eso. Es decir: lo ideal para uno, o lo que uno podría hacer, es algo desprendido de la experiencia inmediata.' (Kaufman, 2007: 108)

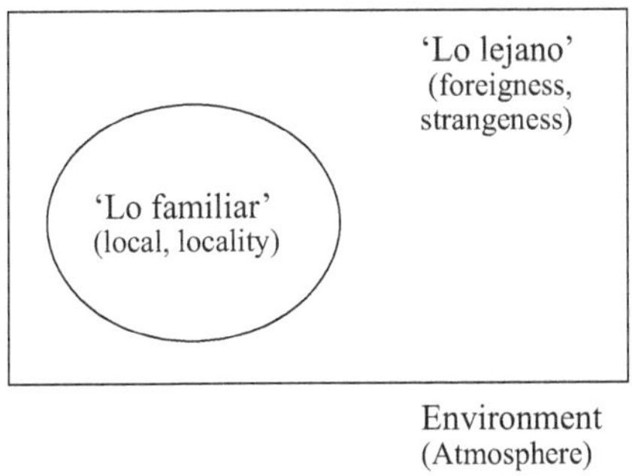

FIGURE 5.2. The idea of Latin America expressed as environment.

As we stated in Chapter 4, environment was conceived/envisaged as a form of space, or more precisely, as the space where the epistemic ground is set (see Table 4.1). Environment in this sense can be understood as *atmosphere*, which is expressed/conceived as local (familiar) and/or foreign (distant, far away). The notions of 'ambiente', 'escenario' and 'amueblamiento' (see Table 2.3) – along with all the expressions connected to them – are also related to this sense of environment as defined by the familiar and the distant.

3.5 The Generation of Degenerates

Latin America as a space, as a concept and domain, does not generate changes by means of revolutions but by *degradation*. The present questioning of the Western/European tradition, contrary to what many specialists affirm (see for example Hardt/Negri, 2000; and Wallerstein, 2004), is not a result of any particular political, ideological or intellectual operation or strategy. It is a simple and natural local evolution – in the Maturana sense of the term (see Maturana, 1995b) – that degrades and degenerates the colonial paradigm imposed by Western/European tradition. Although colonialism can be explained in historiographical terms by using notions of civilization, power and capitalism, those explanations are only a remotely connected with the construction of the actual present. And, paradoxically, the actual questioning of these adverse colonial conditions, has less to do with some critical ideology or historical understanding, and much more to do with a local system that is degraded and has degenerated to the point of not return.

And this degradation is what, in other terms and following Maturana's work, can also be called 'deriva' – the 'deriva' being as *natural* as any cultural and intellectual element can be, that is, irreversible and evolutionary (Maturana, 1995a and 1995b). This 'evolutionism' – culturally, economically, physically – should be considered not by its virtues but by the determinations/results that it produces/generates; and the irreversible aspect should be understood as one of the main elements that compose and determine our actual sense of present. As indicated before, the concepts exist and act independently of the individual will.

We should be concerned more about the understanding of this degradation than about any other ideology, power or army. This condition of *natural degradation* is where one finds the possibilities for 'reversal thinking'. As A. Kaufman, for example, states:

> La nuestra es una cultura desde hace mucho tiempo que no puede pensar la idea de anomalía, accidente, excepción, interferencia; esas nociones no son compartibles con nuestra realidad. En cualquier caso hay una catástrofe, absoluta, apocalíptica, 'todo está mal'. [...] El conocimiento, todo el conocimiento, filosófico, literario, universitario, en la vida moderna está ligado a la guerra. Eso no lo podemos pensar en Argentina. Entonces no podemos hacer la guerra. (Kaufman, 2007: 110 and 117)

'Reversal thinking' tries to overcome these limits of understanding and thereby offer new perspectives on the definition of knowledge. The proposal for the understanding of the local idea of literature, for example, that Julio Ortega made in *Transatlantic Translations* (2006), agrees implicitly with Kaufman's diagnosis and contributes to the direction that we are proposing to take with 'reversal thinking'.

Furthermore, in this general context it is also possible to refer to a sort of *damaged language* to explain the way in which the elements involved in Latin America as space/as a form of space are produced. Therefore the 'reversal perspective' is not trying to do a 'better' analysis than Western/European authors or to 'perform' in relation to a particular subject or matter. This is the crucial point: the 'reversal perspective' assumed a *jump* in the dimension of language, that is, the refusal to implicitly assume the notion of universal language or a universal conceptual system ('science', 'Spanish', etc.). And of course this rejection should work in two ways: (i) in the same way that any universal language is rejected in relation to Western/European traditions, (ii) equally any attempt to establish universal ideas about Latin America, as some authors still do (see for example Galeano, 1997 and 2001), should be avoided for exactly the same reason: epistemic simplification. This is why it is sometimes quite difficult to discuss Latin America as a concept, because there is a long tradition where universalism, in philosophical and linguistic terms, was confused with the defence of some sort of local identity. As already mentioned (Chapter 3), that was another *turn* in the Western/European colonialism implemented at a local level.

The universal assumptions, within the Western/European tradition, are no longer operational. The Western/European framework (enlightenment, modernity, etc.) has not disappeared but its *heuristic power* has another status. Instead of operating as a contemporary sociological condition and/or conceptual category it has become a sort of historical phenomenon or, even worse, a kind of historicism that affects all intellectual activity. This is also why the 'reversal perspective' is not simply about interpretations or finding hermeneutic holes in European-generated theories, but it is about thinking spatially, as we have already stated in Chapter 1. And this is also why, instead of focusing on 'being' ('ser y tiempo', 'temporalidad'), we should focus on that which does not include 'time' as a relevant and constitutive characteristic or category of understanding, such as, the notion of *cosa* that we presented in Chapter 1.

Finally, why should the reader turn his/her attention to a hypothetical idea of 'reversal thinking' when other more tested critical ideas already have their own historiography in relation to the concept of Latin America? The point is not to abandon those useful sets of ideas, like the 'theology of liberation' or the perspective of 'border thinking', but to examine them under the light of two conditions: first, to look at them *from the other side* or create another side for them *another side* (see Figure 5.1, and its respective comments); second, to interpret them from a perspective where the boundaries of what is considered as social no longer exist.[14] If these two perspectives are implemented, we can improve a set of ideas like the 'theology of liberation'. With the first action those sets of ideas will become more connected to the present and/or to the construction of what we understand as present, while with the second action the sets of ideas will be able to face the increasingly sophisticated and non-visible dimension of colonialism in a much more efficient way.

We are not so much degenerated by choice as degraded/corrupted by circumstances that are outside of our control and, in same cases, outside our immediate understanding. 'Reversal thinking' is a perspective that acknowledges this condition and is used as a base and starting point. Conspiracy theories about history and social victimization have ruined many opportunities for effective knowledge within Latin America. Moreover, they have prevented the construction of a credible and efficient sense of locality. It would no doubt be cool and trendy to consider ourselves in a libertarian position regarding Western/European colonialism, but this statement of reference – or any other – will be useless if it does not go together with a conceptual ground that not only endeavours to name what is considered as 'familiar' but also, and in many cases even more importantly, to establish/determine what is understood as 'non-familiar'.

4. Bibliography

Abraham, Tomás. *El último Foucault* (Buenos Aires: Sudamericana, 2002).

Arias, Santa/Melendez, Mariselle (eds.). *Mapping Colonial Spanish America. Places and Commonplaces of Identity, Culture and Experience* (Lewisburg: Bucknell University Press, 2002).

14 'No me importa que las sociedades estén unificadas o alguna cosa por el estilo; no es que quiero eso. Pero me parece que si uno no entiende el modo en que eso acontece en las sociedades como fenómenos colectivos, si uno no toma ese problema como problema, no puede entender lo que nos ocurre.' (Kaufman, 2007: 116).

Allen, Richard. *Projecting Illusion. Film Spectatorship and the Impression of Reality* (Cambridge: Cambridge University Press, 1997).
Barrett, Rafael. *El dolor paraguayo* (Caracas: Biblioteca Ayacucho, 1978).
Benveniste, Emile. *Essays de linguistique generale* (Paris: Gallimard, 1966).
Beorlegui, Carlos. *Historia del pensamiento filosófico latinoamericano* (Bilbao: Universidad de Deusto, 2006).
Bevilacqua, Piero. *L'utilità della storia. Il passato e gli altri mondi possibili* (Rome: Donzelli, 2007).
Boltanski, Luc/Chiapello, Ève. *De la justification. Les économie de la grandeur* (Paris: Gallimard, 1991).
———. *Le nouvel esprit du capitalism* (Paris: Gallimard, 1999).
Bourdieu, Pierre. *Ce que parler veut dire* (Paris: Fayard, 1982).
———. *Homo Academicus* (Paris: Minuit, 1984).
———. *Les règles de l'art* (Paris: Seuil/Libre Examen, 1992).
Brotherston, Gordon. *Image of the New World. The American Continent portrayed in Native Texts* (London: Thames and Hudson, 1979).
———. *Book of the Fourth World: Reading the Native Americas through their Literature* (London: Cambridge University Press, 1992).
———. *Painted Books from Mexico. Codices in the United Kingdom collections and the World they represent* (London: British Museum, 1995).
———. *Footprints through Time: Mexican Pictorial Manuscripts* (Bloomington: Lilly Library, 1997).
Burke, Peter. *Eyewitnessing. The Uses of Images as Historical Evidence* (London: Reaktion, 2001).
Campa, Román de la. *Latin Americanism* (Minneapolis: Minnesota University Press, 1999).
Canaparo, Claudio. *The Manufacture of an Author* (London: King's College London, 2000).
———. *Ciencia y escritura* (Buenos Aires: Zibaldone: 2003)
———. *Muerte y transfiguración de la cultura rioplatense* (Buenos Aires: Zibaldone, 2005).
———. 'La consumación del realismo', in G. Fabry/C. Canaparo, eds., *El enigma de lo real. Las fronteras del realismo en la narrativa del siglo XX* (Bern: Peter Lang, 2007), pp. 199–275.
————. *El imaginario Patagonia. Ensayo acerca de la evolución conceptual del espacio* (forthcoming).
Carr, P./Pond, G. *Second Life. Guida turistica essenziale* (Milan: Isbn Edizioni, 2007).
Casey, Edward S. *Getting Back into Place. Toward a Renewed Understanding of the Place-World* (Bloomington: Indiana University Press, 1993).
———. *The Fate of Place. A Philosophical History* (Berkeley: University of California Press, 1997).
Cassi, Aldo Andrea. *Ultramar. L'invenzione europea del Nuovo Mondo* (Bari: Laterza, 2007).

Caturelli, Alberto. *Historia de la filosofía en la argentina 1600–2000* (Buenos Aires: Ciudad Argentina/Universidad del Salvador, 2001).
Choay, Françoise. *Pour une anthropologie de l'espace* (Paris: Seuil, 2006).
Clam, Jean. *Trajectoires de l'immatériel. Contribution à une théorie de la valeur et de sa dématérialisation* (Paris: CNRS Editions, 2004).
Correas, Carlos. *La operación Masotta* (Buenos Aires: Catálogo, 1991).
Deleuze, Gilles. *Cinema 1. L'image-mouvement* (Paris: Minuit, 1983).
———. *Cinema 2. L'image temps* (Paris: Minuit, 1985).
Del Sarto, Ana/Ríos, Alicia/Trigo, Abril (eds.). *The Latin American Cultural Studies Reader* (Durham: Duke University Press, 2004).
Dussel, Enrique. *Philosophy of Liberation* (Maryknoll: Orbis Books, 1985).
Elias, Norbert. *La civilisation des moeurs* (Paris: Calmann-Lévy, 1973).
———. *La dynamique de l'Occidente* (Paris: Calmann-Lévy, 1975).
Fernández, Silvia. 'The Origins of Design Education in Latin America: From hgf in Ulm to Globalization', in *DesignIssues*, Winter 2006, vol. 22, num. 1, pp. 3–19.
Frampton, Daniel. *Filmosophy* (London: Wall Flower Press, 2006).
Freire, Paulo. *Pedagogy of the Oppressed* (London: Penguin, 1993).
Foerster, Heinz von. *Cybernetics of Cybernetics* (Urbana: University of Illinois, 1974).
———. *Understanding Understanding: Essays on Cybernetics and Cognition* (New York: Springer-Verlag, 2003)
Foucault, Michael. *L'archéologie du savoir* (Paris: Gallimard, 1969).
———. *Les mots et les choses* (Paris: Gallimard, 1966).
———. *Scritti letterari* (Milan: Feltrinelli, 1971).
Galeano, Eduardo. *Open Veins of Latin America: Five Centuries of the Pillage of a Continent* (New York: Monthly Review Press, 1997).
———. *Upside Down: A Primer for the Looking-glass World* (London: Picador, 2001).
Ginzburg, Carlo. *Wooden Eyes. Nine Reflections on Distance* (London: Verso, 2002).
González, Horacio. *Restos pampeanos. Ciencia, ensayo y política en la cultura argentina del siglo XX* (Buenos Aires: Colihue, 1999).
Gracia, Jorge J. E. *Hispanic/Latino Identity. A Philosophical Perspective* (London: Blackwell, 2000).
Grosfoguel, R./Maldonado-Torres, N./Saldívar, J. D. (eds.). *Latino/as in the World-system* (St. Paul: Paradigm, 2005).
Gutiérrez, Gustavo. *A Theology of Liberation* (Maryknoll: Orbis Books, 2005).
Hart, Stephen. '"El Oficio de escribir": Some Notes on Literary Print Culture in Spanish America in the Twentieth Century', in *Neophilologus*, vol. 83, num. 3, 1999, pp. 387–409.
Herrenschmidt, Clarisse. *Les trois écritures. Langue, nombre, code* (Paris: Gallimard, 2007).
Huntington, Samuel P. *The Clash of Civilizations* (London: Free Press, 2002).
Ingaramo, Elias. *Turbios purgatorios* (Rosario: Ideacción, 1992).

Kagan, Richard (ed.). *Spain in America. The Origins of Hispanism in the United States* (Urbana: University of Illinois, 2002).
Kaufman, Alejandro. 'Ontología del presente', in *El río son orillas*, 102/103 (2007), Buenos Aires, pp. 102–123.
Keating, Ana Louise (ed.). *EntreMundos/AmongWorlds: New Perspectives on Gloria Anzaldúa* (New York: Palgrave MacMillan, 2005).
Keating, Ana Louise. *Women Reading, Women Writing: Self-Invention in Paula Gunn Allen, Gloria Anzaldúa and Audre Lorde* (Philadelphia: Temple UP, 1996).
Knorr Cetina, Karin. *Epistemic Cultures. How the Sciences Make Knowledge* (Cambridge, Mass.: Harvard University Press, 1999).
Kusch, Rodolfo. *El pensamiento indígena y popular en América* (Buenos Aires: Hachette, 1977).
Las Casas, Bartolomé de. *La controverse entre Las Casas et Sepúlveda* (Paris: Vrin, 2007).
Latour, Burno. *Petites leçons de sociologie des sciences* (Paris; La Découverte, 1993).
———. *Reassembling the Social. An Introduction to Actor-Network-Theory* (Oxford: Oxford University Press, 2005).
Latour, Bruno/Weibel, Meter (eds.). *Making Things Public. Atmosphere of Democracy* (Cambridge, Mass.: MIT Press/ZKM, 2005).
Levene, Ricardo. *Historia de las ideas sociales argentinas* (Buenos Aires: Espasa Calpe, 1947).
Lindqvist, Sven. *Terra Nullius* (Paris: Éditions des Arènes, 2007).
Martín-Barbero, Jesús. *Al sur de la modernidad. Comunicación. Globalización y multiculturalidad* (Pittsburgh: Instituto Internacional de Literatura iberoamericana, 2001)
Martínez Estrada, Ezequiel. *Muerte y transfiguración de Martín Fierro* (Mexico: FCE, 1948).
———. *La cabeza de Goliath. Microfísica de Buenos Aires* (Buenos Aires: Nova, 1957).
———. *Análisis functional de la cultura* (Buenos Aires: CEAL, 1967).
———. *Sarmiento* (Buenos Aires: Sudamerica, 1969).
Maturana, Humberto. *La realidad: ¿Objetiva o construida?. I. Los fundamentos biológicos de la realidad* (Barcelona: Iberoamericana/Anthropos, 1995). [1995a]
———. *La realidad: ¿Objetiva o construida?. II. Los fundamentos biológicos del conocimiento* (Barcelona: Iberoamericana/Anthropos, 1995). [1995b]
McKenzie, Donald. *Bibliography and Sociology of Texts* (London: British Library, 1986).
Mendes, Candido. *Le défi de la difference. Entretiens sur la latinité* (Paris: Albin Michel, 2006).
Mignolo, Walter. 'Signs and Their Transmission: The Question of the Book in the New World', in Hill Boone, Elizabeth/Mignolo.W. D. (eds.), *Writing Without Words* (Durham: Duke University Press, 1994), pp. 220–270. [1994a]

———. 'Afterword: Writing and Recorded Knowledge in Colonial and Poscolonial Situations', in Hill Boone, Elizabeth/Mignolo.W. D. (eds.), *Writing Without Words* (Durham: Duke University Press, 1994), pp. 292–310. [1994b]

———. 'Espacios geográficos y localizaciones epistemológicas: la ratio entre localización geográfica y la subalternización de conocimientos', en *Pensar* [Bogotá], número 34, available at www.javierana.edu.co/pensar/Rev34. [1996]

———. *Local Histories/Global Designs. Coloniality, Subaltern Knowledge and Border Thinking* (Princeton: Princeton University Press, 2000). [2000a]

———. 'The Many Faces of Cosmo-polis: Border Thinking and Critical Cosmopolitanism', in *Public Culture*, vol. 13, num. 3, 2000. [2000b]

———. *The Idea of Latin America*, London: Barckwell, 2005.

Mignolo, W./Schiwy, F. 'Double Translation', in Maranaho, T./Strek, B. (eds.). *Translation and Ethnography. The Anthropological Challenge or Intercultural Understanding* (Tucson: University of Arizona Press, 2003), pp. 3–29.

Moraga, Cherríe. *The Last Generation: Prose and Poetry* (Boston: South End Press, 1993).

Moraña, Mabel (ed.). *Nuevas perspectivas desde/sobre América Latina. El desafío de los estudios culturales* (Pittsburgh: Instituto Internacional de Literatura iberoamericana, 2000).

Navvaro García, J. Raúl (coord.). *Literatura y pensamiento en América Latina* (Seville: Escuela de estudios Hispano-Americanos/CSIC, 1999).

Hardt, Michael/Negri, Antonio. *Empire* (Cambridge, Mass.: Harvard University Press, 2000).

Ortega, Julio. *Transatlantic Translations. Dialogues in Latin America Literature* (London: Reaktion, 2006).

Pauli, Lori. *Manufactured Landscapes. The Photographs of Edward Burtynsky* (Ottawa/New Haven: National Gallery of Canada and Yale University Press, 2006).

Petrucci, Armando. *Alfabetismo, escritura, sociedad* (Barcelona: Gedisa, 1999).

Piaget, Jean. *L'épistémologie génétique* (Paris: PUF, 1970).

Pietromarchi, Bartolomeo (ed.). *The [Un]common Place. Art, Public Space and Urban Aesthetics in Europe* (Barcelona: Actar, 2005).

Pineda Franco, Adela/Sánchez Prado, Ignacio (eds.). *Alfonso Reyes y los estudios latinoamericanos* (Pittsburgh: Instituto Internacional de Literatura Iberoamericana, 2004).

Prebisch, Raúl. *Hacia una dinámica del desarrollo latinoamericana* (Mexico: FCE, 1963).

Rama, Ángel. *Transculturación narrativa en América Latina* (Mexico: Siglo XXI, 1982).

———. *La ciudad letrada* (Hannover: Ediciones del Norte, 1984).

Ronell, Avital. *American philo* (Paris: Stock, 2006).

Sarlo, Beatriz. *La batalla de las ideas (1943–1973)* (Buenos Aires: Ariel, 2001).

Sennett, Richard. *The Culture of the New Capitalism* (New Haven: Yale University Press, 2006).

Senté, Richard. *The Culture of the New Capitalism* (New Haven: Yale University Press, 2006).

Schutte, Ofelia. *Cultural Identity and Social Liberation in Latin American Thought* (Albany: SUNY Press, 1993).

Simon, Josef. *Signe et interprétation* (Villeneuve d'Asq: Presses Universitaires du Septentrion, 2004).

Spitta, Silvia. *Between Two Waters. Narratives of Transculturation in Latin America* (College Station: Texas A & M University Presses, 2006).

Vattimo, G./Rovatti, P. A. (eds.). *Il pensiero debole* (Milan: Feltrinelli, 1983).

Wallerstein, Immanuel. *Geopolitics and Geoculture. Essays on the Changing World-System* (Cambridge: Cambridge University Press/Maison des Sciences de l'Homme, 1991).

———. *The End of the World as We Know It. Social Science for the Twenty-First Century* (Minneapolis: Minnesota University Press, 1999).

———. *The Limits of Nineteenth-Century Paradigms. Unthinking Social Sciences* (Philadelphia: Temple University Press, 2001).

———. *World-System Analysis* (Durham: Duke University Press, 2004).

White, Hayden. *Tropics of Discourse* (Baltimore: Johns Hopkins, 1978).

———. *The Content of The Form* (Baltimore: Johns Hopkins, 1987).

Zabunyan, Dork. *Gilles Deleuze. Voir, parler, penser au risque du cinéma* (Paris: Presses Sorbonne Nouvelle, 2006).

Zea, Leopoldo. *América Latina en sus ideas* (Mexico: Siglo XXI/UNESCO, 1993).

Postface
The Wasted Land

> I have lived in this country [UK] far longer than in Bavaria, but reading in English I become self-conscious about having a funny accent. Unlike Conrad or Nabokov, I didn't have circumstances which would have coerced me out of my native tongue altogether. But the time may come when my German resources begin to shrink. It is a sore point, because you do have advantages if you have access to more than one language. You also have problems, because on bad days you don't trust yourself, either in your first or your second language, and so you feel like a complete halfwit.
>
> W. G. Sebald (2001: 4)

Latin America as an idea is a place that, like thousands of other locations, exists in many forms and versions. However, Latin America is perhaps the first place on this planet consistently *constructed towards the future*, from the possibilities of what could happen or, more recently, from expectations about the future. Nothing makes Latin America more distinctive than this existence in the future, in the distance or, to be more precise, in a place that in order to be considered as *here* needs to be elsewhere. This is the main paradox that this book was really trying to explore.

Readers might question the implicit assumption of this work which is that Latin America is considered as a relevant subject and, more specifically, as a *leading case* in contemporary thought. We are well prepared to bring in arguments about population, vegetal life or other classical aspects that can prove this relevance. Instead, we prefer to insist on this *unique* condition of Latin America as a concept: the fact that it was constantly built towards the future and therefore its meanings come from this regular connection to things, objects and circumstances that have not yet happened.[1] The fact that Latin America – as a continent

1 Of course, Latin America is an invention, a rhetorical artefact. However this condition does not make it any less effective or relevant as a notion, as an

or as a region – has always had a *privileged* relationship with Western/European colonial perspectives is also something that might explain this *uniqueness* (see for example Mignolo, 2005: 22–34).

Many authors dedicated to the study of Latin America believe that Latin America – as a concept, as a culture, or as history – constitutes a *subject* and, therefore, when they try to make distinctions and criticism, everything reduces to the question of the content of the arguments. Aware of this situation, other authors have recently decided to work from a different and changed perspective. As we have mentioned already, E. Dussel and W. Mignolo feature among this group of authors who, while exploring Latin America as concept, have tried to generate a more appropriate epistemic discussion.[2] However, under certain circumstances Dussel (1977) and Mignolo (2005) also give privilege to arguments according to their *content*, especially when a normative view or a prospective interpretation is proposed.

According to our hypothesis, the distinction that could explain many particularities and, no less importantly, that enables us to establish a local and independent *horizon d'attente*, is located at the epistemological level rather than at the level of the contents. Because of this condition, the origin of the ideas (country, university, etc.) or the origin of the authors ('patronímico') is less relevant than the role and function that the ideas and the authors play in relation to the environment or the market. In this sense I believe, for example, that the so-called radical constructivism in Western philosophy is more liberating to the concept and idea of Latin America than the popular 'teoría de la liberación'. The response to the local problems is not placed in particular subjects or contents but in the construction of a way of thinking, in the form in

imaginary location, as a homeland, as a semantic unit or as a place-to-be. When what is considered as natural is created and when the geography of a place becomes a cartographic phenomenon, to mention just two epistemic aspects of the transcultural condition (see Canaparo, forthcoming), then the fact that Latin America was in first instance a narrative entity makes it no different from the majority of States and Nations created during the nineteenth century.

2 W. Mignolo for example refers a 'search to another logic' (2000: 47 ff.), as a way to establish an *outside* point of view to the Western/European tradition and culture.

which we construct a way of thinking. Of course, authors like P. Freire, G. Gutiérrez and E. Dussel, to mention just a few of those related to the above-mentioned 'teoría de la liberación', are superb sources of ideas and perspectives, but in context, in the domain in which their ideas circulate, they become less effective than predicted. And other non-local generated perspectives have perhaps a greater long-term effect without necessarily being colonialist approaches.

The proposed 'reversal thinking' tries to work on precisely this point: the current level of complexity in local and intellectual issues has made the content-based perspective ('contenidos', 'tramas') obsolete. It is impossible nowadays to conceive an issue ('problemática') in isolated terms.[3] This is even more relevant in Latin America – as a ground or as a concept – because it has always been envisaged as being part of the Western/European domain, despite at the same time being understood as a 'continent' (see for example Mignolo, 2005: 22–34 and 128–135).

It is an unfortunate folkloric misunderstanding to think that interpersonal conditions of communications in Latin America are 'special' and share nothing in common with what happens elsewhere, including the European territory. The specificity of Latin America as a cultural space, as we have tried to demonstrate here, is not defined by this folkloric stand but by the way in which perception and knowledge is constructed. If we focus on what people do, then the Western/European paradigm emerges; if we focus on how people do things, then a variety of alternatives are available. The matter of this reflection is then less about what people do than how they do it.[4] Equally, to focus on what authors say is not as useful as focussing on exploring why, how and from where authors say what they say.[5] The perspective of 'loci of enunciation', as

3 The so-called 'complexity theory', named and developed mainly from the works of Heinz von Foerster, that we subsume under the title of radical constructivism, is certainly pertinent here (see for example Andreewsky/Delorme, 2006).

4 From this proposition we can produce a reading of the limits of all neo-Marxist theories which were predominant among Latin America's intellectual communities during the twentieth century.

5 This situation introduced two issues that were not extensively considered by local thinkers during the twentieth century: (i) the fact that the present is a very

addressed by Mignolo (2000a and 2005), is a possible way forward in the matter of what is understood nowadays as Latin America. However we also believe that Maturana's – and others' – radical constructivism has been more influential and radical in this way (see for example Maturana, 1995a and 1995b; and Maturana/Varela, 1972).

There is an 'ingenuidad de la vida diaria', as Maturana (1995a: 75) refers to it, that is probably the worst form of ignorance concerning Latin America as analyzed here. We refer to the assumed view that concepts and meanings are constructed from pure information and that it is their change and/or modification that makes things, objects and circumstances similar or different in terms of knowledge. This book points to a different conclusion: information is an irrelevant element of knowledge and thinking as understood here. Only an *epistemological perspective* can establish an efficient distinction between things and individuals.

The original idea for this work, besides what has been said already, came from the need for *thinking from nowhere*, as this 'nowhere' is my 'mother land', my *hinterland.* And by this I am not referring simply to bibliographic or historiographic authorities, but also to topographies, imaginary places, and the everyday language. Of course I have discussed books, imperial ideas and authors, but they were never at the forefront point of my thinking. This is a book about ideas and politics but it is written in a *poiesis form* (Maturana/Varela, 1972 and 1980), that is, focusing mainly on the construction and development of two issues: conceptual semantic accuracy and a general sense of language.

---

complex category which is almost impossible to sustain in one single concept; and (ii) the fact that, in texts as in the media, it is no longer possible to have *one voice* (or many single voices) but instead one finds many different circumstances of enunciation and/or conditions of writing or speech. On observing how obsolete the local literary theory has become one obtains a clear understanding of the consequences of ignoring this complexity.

1. The Place of Knowledge

Except for the works of Enrique Dussel and, to a certain extent those of Edmundo O'Gorman (1906–1995), among the few authors working within a similar conceptual framework, there is, from an epistemic perspective, no relation between the works of local authors that have dealt with Latin America understood as subject or idea, and the *concept* of Latin America as discussed here. It is for this reason that, contrary to tradition, we have not based our arguments on the history or the historiography of the idea of Latin America as it has been developed by native authors or foreign academics.

This situation is connected with what we call the location of knowledge, that is to say, (i) how knowledge becomes a local element, (ii) how what is local is defined by an idea of knowledge, and (iii) how the development of a 'locality' and 'localization' constantly needs to be referred to something described and understood as knowledge. Therefore our choice was not only about some historiographical local approach but was also related to the way in which the relation between knowledge and what is understood as local has been developed by native authors. Briefly, we have been pushing in six directions:

1. The role of science or scientific knowledge at the local level is determined by a classical colonial approach to learning and understanding, and the local system of authority and education was based on this (on this topic see for example Peset, 1987).
2. Although some aspects of the classical 'colonial matrix of power' (Mignolo, 2005: 4) remain, the recent changes at the level of writing, media development and definition of space, bring another new form of colonialism, so dear to many young local authors, that generally speaking we can identify as *brand culture* (on this subject see for example Schroeder/Salzer-Mörling, 2006).
3. At the core of the discussion about new forms of colonialism, new formulations of the Western/European culture domination and new authoritative forms of knowledge, there is a not-very-popular debate about thinking as activity grounded mostly in

an approach to space, and also as a form of contrast with the classical philosophical understand in the Western/European domain (on this point see for example Naficy, 1999).

4. Even in the Western/European territory the activity of writing is nowadays starting to be considered in relation to a 'visuality', that is, writing has become more a semiotic domain and less an alphabetically orientated linguistic activity (on this subject see for example Westphal, 2007).
5. The sociological conceptual systems which originated historically within the Western/European paradigm of knowledge have started to come into conflict (of interpretation, of meaning, of efficacy) and are losing their validity and legitimacy because of the impact of the visual in the constitution of communities and their 'localization' (on this subject see for example Macé, 2006).
6. The loss of a clear hierarchy of values in speculative terms makes the activity of thinking, on one hand, something more open and uncertain, but on the other hand, more corporate/corporately orientated and closely connected with the development of institutions. Another consequence of this lack of dominant values is the fact that all new subject areas that become commercially dominating affect the thinking. Tourism, for example, is one of those areas (on this topic see for example Rojek/Urry, 2000).[6]

These directions were considered in the light of Latin America viewed as a concept, but they can also be used in reference to other spatial areas and concepts. What we consider to be *unique* in the case of Latin America is the way in which *thinking* becomes a natural local activity by degrading or deforming what is understood in Western/European terms as 'philosophy'.

6 The recent interest of local authors in travelling, remote locations, and transculturization, understood as a close exchange with Europe and United States, is indeed part of this tendency. Another result of this condition is the increasing relevance that cultural *événements* originating in Europe/US acquired in local terms.

In this context, this book aspires to contribute to one of the main problems in 'proving' the existence and functionality of 'Latin American thought', that is, the issue of *sustainable knowledge*. As we have already stated, the historiographic expansion and domination of the Western/European culture prevents the extreme of any/equal historiography, therefore establishing a sort of *sustainable knowledge* with regard to any form of 'Latin American thought' implies bringing the Western/European principle of historiography into question. A 'paradigm of conviviality' as Mignolo calls it, is the minimum request here (see Mignolo, 2005: 143 and 161).

Finally, another aspect that this book tries to address is the fact that attempting to think from the margins and borders of dominant tendencies does not mean refusing to engage with the Western/European cultural products and, especially, with their historiographical principles and conditions and with their conceptual domain. The idea of *reversal thinking* was partially proposed to make this perspective possible. Other current proposals, as we have already seen, work according to a similar pattern: W. Mignolo's previously mentioned notion of 'border thinking' is probably one of the most notorious examples in the academic sphere (Mignolo, 2005).

2. 'Ética del resentimiento'

Many books written on similar subjects or issues to the one we are dealing with here – the idea/concept of Latin America, for instance – are delivered in what I will provisionally call a *celebratory model*. That is, their authors explain how much they owe to good conversation with academic colleagues and students, to the very auspicious financial conditions of the leading universities they are working with, to the many seminars and symposiums they have attended and, last but not least, to the supportive family they surely have. On the contrary, this book came from the opposite direction and environment. This book is an 'ejercicio de sombras', to quote the German philosopher from Röcken. It comes from the margins of Western/European academia and social life. And furthermore, even if only partially, it is because of those conditions that it was written. There is only a tiny distinction here between radical thinking and 'resentimiento social' (social resentment).

Alexis de Tocqueville (1805–1859) predicted in the nineteenth century that the tendency and ultimate result of democracy was individual and social *ressentiment*. Since then many authors have been exploring what has become a principal reality in today's social life (see for example Sennett, 1978). The university environment, especially in humanities and literary studies, is a corporative domain where this democratic condition has been developed quite well – to such an extent that some critics label it 'School of Resentment' (see Bloom, 1994). More and more the corporations that we still recognize as universities have become structures of social promotion or, in the best of cases, institutions of social contention. The university as a place of intellectual development no longer exists, it has been transformed. As a result, the colonialism exercised by them is no longer based on meritocracy, training or language but on bureaucratic accomplishment and management behaviour. The new university authorities call this *by the book* obedience a 'good career'. As ever the universities have become an appendix of the State's policy on social issues. Paradoxically, just as in society, the more open to society a university is, the greater the resentment it produces. And because of this there is a sort of Weberian 'ethic of resentment' ('ética del resentimiento') that explains the *esprit* of the knowledge in Western/European domains.

How this 'resentimiento' works and what consequences it has for the constitution of an idea/concept of Latin America, are just two of the many questions that this work has been attempting to address.

3. 'Teoría del terreno baldío'

The *displacement* of Latin American authors and academics working in central/imperial institutions is not a new phenomenon and can even be traced back to the colonial period, in such cases as the previously mentioned Inca Garcilaso, and many other authors or intellectuals, that have focused at least part of their studies on Western/European domains and institutions. Latin American authors and travelling academics normally find their way into Western/European institutions by using/profiting from the *folkloric* element that is attached to them – such as language, skin colour, particular body skills, etc. Some of them – probably the most commercially successful ones – used it as a passport to other areas or disciplines, others used it to launch their careers. This book was also written out of this limitation: Latin American thought does not exist in terms of 'philosophy' within Western/European universities and academic institutions. Therefore, those who consider themselves to be working on 'Latin American thought' are normally forced to work in the domains of literature, cultural analysis or history, areas which are more permeable to the *folkloric* element.

And this *displacement* is not just unique to Latin American authors, it also happens to authors whose work is connected to Africa or Asia and, moreover, also occurs within – and between – Western/European institutions for obvious marketing and financial reasons (see for example Bourdieu, 1989; Becher, 1989; and Jacob/Hellström, 2000). This book has also been written outside of this *alienated experience*, out of the fact that it makes no difference what you do, you will always be a *foreigner* in terms of knowledge or, more precisely and to quote Mignolo, in terms of *gnosis* (see Mignolo, 2000a).

Of course, in the past many local authors tried to confront the colonialist attitude that considers Latin American as a savage and uncivilized place. These authors have also tried to combat this and work towards a 'reivindicación' of Latin America as decent land, full of potential opportunities for development. This book, along with the 'reversal thinking' theory that it contains, points in quite the opposite direction: for us Latin America, as an historical stand, is a 'wasted land', therefore, rather than confronting once again the above-mentioned colonialism, we embrace it.

Instead of being concerned about the hypothetical complacency of this strategy, we believe the most effective way to confront colonialism is by processing it, distorting it, twisting it, using it – a *use* that happens anyway, as we have seen throughout this book. And this is why we disagree, once again, with the *celebratory model* of colonialism that is behind of many of those classical proposals – for example, by authors like Isabel Allende (1942–) (*Mi país inventado*, 2003), Edward W. Said (*Out of Place*, 1999), Salman Rushdie (1947–) (*Imaginary Homelands*,

1991), and Tzvetan Todorov (1939–) (*L'homme dépaysé*, 1996), to mention just few well-known examples. On the contrary, this book shares the experiences and views of those people who envisage colonialism as a conceptual and personal problem and who, as a consequence of their views, do not explore colonialism using the traditional and simplistic terminology of academia or popular culture. I am thinking of a wide range of authors and artists including – to give just a few examples of written works – Lascano Tegui (1887–1966) (*Muchacho de San Telmo*, 1895), Ariel Dorfman (1942–) (*Heading South, Looking North*, 1998), Elias Ingaramo (1944–1996) (*Constitución*, 1994), Isabel Fonseca (1962–) (*Bury Me Standing*, 1996), Albalucía Angel (1939–) (*Estaba la pájara pinta sentada en el verde limón*, 1975), Italo Calvino (*Eremita a Parigi*, 1994), Arnaldo Calveyra (1929–) (*El hombre de Luxemburgo*, 1997; *Diario del fumigador de guardia*, 2002). And also in the public standing of authors such as Alejandro Kaufman (2008) in Argentina or Jorge Drexler (1964–) (2007) in Uruguay, among others. Of course, notorious authors like Gloria Anzaldúa or Juan Gelman (1930–), to name two examples, are very useful indeed but they are very well-known already and they also present another historiographic problem: their public notoriety and visibility within Western/European academia and market – and, of course, the fact that these authors are not dealing/considering with this *public standing* of their work.

Working in Western/European academies in the field of Latin America is like working in a 'wasted land', in a space that is not relevant to the main conceptual and financial systems of universities. Furthermore, working on Latin America outside the aforementioned *celebratory model* is also being in a space that is considered as 'wasted land'. Finally, as we have stated elsewhere (see Canaparo, 2005: 121–148 and 221–322), it is almost impossible to think about Latin America in terms of knowledge – as a culture, a concept or as a history – without considering the notion of 'chacarita' ('chatarra', 'deshecho', 'basura'), that is to say wasted materials, objects and people. These are the three ways in which the dimension of 'wasted' appears and justifies, in our view, the central role that it has in defining local knowledge and in taking a position with regard to colonial perspectives. It is unfortunate, for example, that almost no-one in Latin America has picked up on Rogério Sganzerla's (1946–2004) brilliant perspective on the 'aesthetics of garbage' or tried to develop it in knowledge terms (see for example R. Johnson/R. Stam, 1995; and also G. Rocha, 1981).

It is also from this standpoint that the 'reversal thinking' approach tries to work. Within the Western/European context 'philosophy' had been considered as the 'science of sciences' in cultural and conceptual terms. More recently, and in connection with the monopolistic dominance of the scientific paradigm, understood as a major way of thinking, 'epistemology' has begun to cover the classical role assigned to 'philosophy' within the Western/European domain.

Many authors have discussed the incompatibility between the Western/European notions of 'philosophy' and 'epistemology' and the Latin American space (see for example Dussel, 1994). This book is not innovative in this sense as, assuming this situation, it is attempting to figure out what, in the Latin American context, provides the 'función agonal del pensamiento'. In other words, it attempts to say what, if any, is the gravitational point within the Latin American domain. Furthermore, considering the fact that there does not seem to be one *single gravitational point* but many, our task has been to define the sense of the local – or locality – as a way to establish the *core* of thinking. It has been very useful to observe how and why this 'locality' could be connected to the *disengagement* ('delinking') that authors like W. Mignolo (2005) propose as an intellectual methodology. However, our proposal points in the opposite direction as we have postulated that Western/European perspectives and concepts are *degraded* ('degradación') by overuse ('haciendo uso extremo', 'interpretación extrema'). In this sense, there is no *outside* perspective from which we can think/consider the local or, more precisely, *what is local*. The idea of the 'wasted' is also related also to this wish for degradation and non-independence.

4. 'El país de no me acuerdo'

It is this regular perspective about what is not there but is required to conform to the sense of the present that, as we have already mentioned, generates the need to construct imaginary homelands, as much in biographical/autobiographical as in conceptual terms. And it is this basic condition of knowledge and understanding that creates the ambivalence towards this *place of origin*: on the one hand we need it, and on the other hand, we would prefer not to have it, to forget it as if it were something that never existed. The historic and troubled relationship with the idea of 'memory', for example, is indeed related to this basic situation – just as it is connected with the situation of 'wasted land' as described before.

It is probably the constraint/restriction generated by this basic condition that has exacerbated the natural local relationship between thinking and seeing. That is to say, it seems to be increasingly evident that that can only be created ('idear') which is seen ('ver', 'visto') or, to be more precise, that can only be developed which can be constructed as seen ('veído', 'visto') – a context in which the local *motto* that 'think is to invent' ('pensar es inventar', 'pensamos inventando') acquires meaning. Edmundo O'Gorman's classic book, *La invención de América*, published in 1958 in Mexico, seems to me one of the best examples of this condition. And if this condition has today become more relevant than ever, more than due to local and already very well-known reasons, it is because with the disappearance of distances (Canaparo, 2005), and the consequent *planetización* of each 'philosophical' question (Wallerstein, 2004), the so-called 'capitalisme cognitif' (Boutang, 2007) has acquired an unexpected local dimension. As *values* become ever more abstract, and as they become ever more 'financially'orientated, so the classical material distribution between 'poor' and 'rich' cultures has less relevance, as the relevant distinction (colonialism) starts to be developed elsewhere.

As the distance between what is understood as local and what is viewed as 'elsewhere' gets shorter, so this basic condition of 'el país de no me acuerdo' (the homeland I cannot remember) becomes more problematic. Instead of having the sensation that this increasing neighbourhood of everything re-values local biographies and concepts, on the contrary, the closeness ('proximidad') makes everything local more 'wasted'. The closer people and terms are to this sort of common world-place (see for example Casey, 1998) the further away people would like to be and the more justification (than ever) concepts require. It is exactly this problematic sense of neighbourhood, for example, applied to contemporary Western/European culture, that Slavoj Žižek has developed in his theory of war, psychoanalysis and neighbourhood (see Žižek, 2007).

However, once again, what seems to be the relevant point here is the fact that, as it becomes more problematic to determine any distance between the local and the faraway, so the local domain seems to be more 'wasted'. And even if this condition is not new, what is new is the fact that any attempt to construct/establish a 'memory', as a defence against this 'wasted' condition, becomes useless – not because of any defect of the notion of memory as such, but because memories have no relevance in Latin America understood as a concept. Quite the contrary, the more 'memory' is placed at the centre of any speculative program or project, as is the case in many local critical attempts, the more dependent on Western/European conceptual colonialism it becomes.

And, as the reader may have already concluded, if Latin America is a place mostly constructed upon the existence of certain *distance(s)*, the present problematic entity of this notion, directly questions the classical conception of Latin America – as a subject, an idea or even as a fixed topography. 'Here' and 'there' have become over-exposed and the reassuring separate dimensions in which they were considered – as was the case of the basic condition of local culture for many years – is no longer available. In this sense, more than ever, Latin America figures paradoxically present at the cultural and intellectual core of the main colonial capitals of the Western/European empires – the recent works of Toni Negri (Hard/Negri, 2000) and Naomi Klein (2000), to mention two very popular cases, are unchallenged proof of this.

Finally, this epistemic situation of knowledge cannot be detached from the entanglement between migration and environment, as we have already mentioned; nor can it be detached from the visual dimension that has become an increasingly dominant factor in education and understanding. The biological constitution of the self, the development of an 'I' on paper ('el yo de papel') and an idea of knowing ('conociendo') are another three elements that, entangled as they are, radically affect our perception of what we determine as local and as knowledge. We have tried to demonstrate that Latin America and the location of knowledge is not so much a convenient and trendy formulation as a dimension that will radically transform our understanding of where we are and where we have come from.

To conclude, I would like to make an historical/historiographical analogy. The fact that all written documents from pre-Colombian civilizations were produced *after* the arrival of the Europeans is just as important as the fact that Castilian was the *vehicular language* that paradoxically allowed the locals to consolidate a corpus of the native languages, like Quechua or Nahuatl. Is not just necessary to state that most of the materials of communication – which were not alphabetically-orientated – were destroyed, it is also relevant to point out that the notions of writing and of *document* were introduced/established by the Europeans and were immediately adopted by the locals to use ('operar', 'funcionar') in their own communities and in the ways they understood what they were doing and what they saw.

I believe that we face a similar situation today with regard to three elements: (i) the generalization of the English language, (ii) the monopolistic association between knowledge and 'science', and (iii) the development of the visual as our major tool for developing a sense of 'world' and 'reality'. As the natives did then, I believe we can today, twist, degenerate and degrade those elements in order to develop our own sense of what is local and what is knowledgeable ('conocible'). 'Reversal thinking' is, I believe, the proposal to initiate this or, at least, one of the many possibilities to start to do so.

5. Bibliography

Allende, Isabel. *Mi país inventado* (Barcelona: Random House/Mondadori, 2003).

Andreewsky, Evelyne/Delorme, Robert. *Seconde cybernétique et complexité* (Paris: L'Harmattan, 2006).

Anzaldúa, Gloaria. *Borderlands/La Frontera. The New Mestiza* (San Francisco: Aunt Lute Books, 1999).

Becher, Tony. *Academic Tribes and Territories. Intellectual enquiry and the cultures of disciplines* (Buckingham: SRHE/Open University, 1989).

Bloom, Harold. *The Western Canon: The Books and School of the Ages* (New York: Harcourt Brace, 1994).

Bourdieu, Pierre. *La noblesse d'état. Grandes écoles et esprit de corps* (Paris: Minuit, 1989).

Boutang, Yann Moulier. *Le capitalisme cognitif* (Paris: Éditions Amsterdam, 2007).

Calveyra, Arnaldo. *El hombre de Luxemburgo* (Barcelona: Tusquets, 1997).

———. *Diario del fumigador de guardia* (Bahía Blanca: Vox, 2002).

Calvino, Italo. *Eremita a Parigi. Pagine autobiografiche* (Milan: Mondadori, 1994).

Canaparo, Claudio. *The Manufacture of an Author* (London: King's College London, 2000).

———. *Ciencia y escritura* (Buenos Aires: Zibaldone: 2003)

———. *Muerte y transfiguración de la cultura rioplatense* (Buenos Aires: Zibaldone, 2005).

———. 'La consumación del realismo', in G. Fabry/C. Canaparo, eds., *El enigma de lo real. Las fronteras del realismo en la narrativa del siglo XX* (Bern: Peter Lang, 2007), pp. 199–275.

————. *El imaginario Patagonia. Ensayo acerca de la evolución conceptual del espacio* (forthcoming).

Casey, Edward S. *Fate of Place: A Philosophical History* (Berkeley: California University Press, 1998).

Dorfman, Ariel. *Heading South, Looking North* (London: Hodder & Stoughton, 1998).

Drexler, Jorge. 'Soy un goloso con la vida', in *Fuera de serie*, número 127, 11 January 2007, pp. 16–17.

Dussel, Enrique. *Filosofía de la liberación* (Mexico: Edicol, 1977).

———. *Historia de la filosofía y filosofía de la liberación* (Colombia: Editorial Nueva América, 1994).

Fonseca, Isabel. *Bury Me Standing* (London: Viking, 1996).

Hardt, Michael/Negri, Antonio. *Empire* (Cambridge, Mass.: Harvard University Press, 2000).

Ingaramo, Elias. *Constitución* (London: La Yapa Ediciones, 1994).

Jacob, Merle/Hellström, Tomas (eds.). *The Future of Knowledge Production in Academia* (Buckingham: SRHE/Open University, 2000).

Johnson, R./Stam, R. (eds). *Brazilian Cinema* (New York: Columbia University Press, 1995).

Kaufman, Alejandro. 'Todos los días nos invaden marcianos', in *Página 12*, Buenos Aires, 13 April 2008.

Klein, Naomi. *No Logo* (London: Picador, 2000).

Macé, Eric. *Les imaginaires médiatiques* (Paris: Éditions Amsterdam, 2006).

Maturana, Humberto. *La realidad: ¿Objetiva o construida?. I. Los fundamentos biológicos de la realidad* (Barcelona: Iberoamericana/Anthropos, 1995). [1995a].

———. *La realidad: ¿Objetiva o construida?. II. Los fundamentos biológicos del conocimiento* (Barcelona: Iberoamericana/Anthropos, 1995). [1995b].

Maturana, Umberto/Varela, Francisco. *De máquinas y seres vivos* (Santiago: Editorial Universitaria, 1972).
———. *Autopoiesis y cognition* (Dordrecht: Kluwer, 1980).
Mignolo, Walter. 'Signs and Their Transmission: The Question of the Book in the New World', in Hill Boone, Elizabeth/Mignolo.W. D. (eds.), *Writing Without Words* (Durham: Duke University Press, 1994), pp. 220–270. [1994a]
———. 'Afterword: Writing and Recorded Knowledge in Colonial and Poscolonial Situations', in Hill Boone, Elizabeth/Mignolo.W. D. (eds.), *Writing Without Words* (Durham: Duke University Press, 1994), pp. 292–310. [1994b]
———. 'Espacios geográficos y localizaciones epistemológicas: la ratio entre localización geográfica y la subalternización de conocimientos', en *Pensar* [Bogotá], número 34, available at www.javierana.edu.co/pensar/Rev34. [1996]
———. *Local Histories/Global Designs. Coloniality, Subaltern Knowledge and Border Thinking* (Princeton: Princeton University Press, 2000). [2000a].
———. 'The Many Faces of Cosmo-polis: Border Thinking and Critical Cosmopolitanism', in *Public Culture*, vol. 13, num. 3, 2000. [2000b]
———. *The Idea of Latin America* (London: Blackwell, 2005).
Naficy, Hamid. *Home, Exile, Homeland. Film, Media, and the Politics of Place* (New York: Routledge, 1999).
O'Gorman, Edmundo. *La invención de América* (Mexico: FCE, 1958).
———. *El proceso de la invención de América* (Mexico: FCE, 2000).
Peset, Luis. *Ciencia y libertad. El papel del científico ante la independencia americana* (Madrid: CSIC, 1987).
Rocha, Glauber. *A revoluçao do Cinema Novo* (Rio de Janeiro: Alhambra/Embrafilme, 1981).
Rojek, Chris/Urry, John. *Touring Cultures. Transformations of Travel and Theory* (London: Routledge, 2000).
Rushdie, Salman. *Imaginary Homelands* (London: Granta, 1991).
Said, Edward W. *Out of Place* (London: Granta, 1999).
Schroeder, Jonathan/Salzer-Mörling, Miriam. *Brand Cultures* (London: Routledge, 2006).
Sebald, G. W. 'The Last Word', in *Guardian Unlimited*, www.guardian.co.uk, December 21, 2001.
Sennett, Richard. *The Fall of Public Man* (New York: Knoft, 1978).
Tegui, Lascano. *Muchacho de San Telmo* (Buenos Aires: Kraft, 1895).
Todorov, Tzvetan. *L'homme dépaysé* (Paris: Seuil, 1996).
Wallerstein, Immanuel. *World-Systems Analysis: An Introduction. Durham* (North Carolina: Duke University Press, 2004).
Westphal, Bertrand. *La géocritique. Réel, fiction, space* (Paris: Minuit, 2007).
Žižek, Slavoj. 'Teme a tu vecino como a tí mismo', conference delivered in the 'Círculo de Bellas Artes', Madrid, 08.03.2007.

General Index

Names Index

List of Figures and Tables

Chapter 1

Chapter 2

Chapter 3

Chapter 4

Chapter 5

Hispanic Studies: Culture and Ideas

Edited by
Claudio Canaparo

This series aims to publish studies in the arts, humanities and social sciences, the main focus of which is the Hispanic World. The series invites proposals with interdisciplinary approaches to Hispanic culture in fields such as history of concepts and ideas, sociology of culture, the evolution of visual arts, the critique of literature, and uses of historiography. It is not confined to a particular historical period.

Monographs as well as collected papers are welcome. Languages of publication are English, Spanish and Spanish-American.

Those interested in contributing to the series are invited to write with either the synopsis of a subject already in typescript or with a detailed project outline to either Dr Claudio Canaparo, Centre for Latin American Studies, School of Arts, Languages and Literatures, University of Exeter, Exeter EX4 4QH, UK, c.canaparo@exeter.ac.uk, or to Hannah Godfrey, Peter Lang Publishing, Evenlode Court, Main Road, Long Hanborough, Witney, Oxfordshire OX29 8SZ, UK, h.godfrey@peterlang.com.

Vol. 1 Antonio Sánchez
Postmodern Spain. A Cultural Analysis of 1980s–1990s Spanish Culture. 220 pages. 2007.
ISBN 978-3-03910-914-2

Vol. 2 Geneviève Fabry and Claudio Canaparo (eds)
El enigma de lo real. Las fronteras del realismo en la narrativa del siglo XX. 275 pages. 2007.
ISBN 978-3-03910-893-0

Vol. 3 William Rowlandson
Reading Lezama's *Paradiso*. 290 pages. 2007.
ISBN 978-3-03910-751-3

Vol. 4 Claudio Canaparo, Fernanda Peñaloza and Jason Wilson (eds)
Patagonia. Myths and Realities. *Forthcoming.*
ISBN 978-3-03910-917-3

Vol. 5 Xon de Ros
Primitivismo y Modernismo. El legado de María Blanchard.
238 pages. 2007.
ISBN 978-3-03910-937-1

Vol. 6 Sergio Plata
Visions of Applied Mathematics. Strategy and Knowledge.
284 pages. 2007.
ISBN 978-3-03910-923-4

Vol. 7 Annick Louis
Borges ante el fascismo. 374 pages. 2007.
ISBN 978-3-03911-005-6

Vol. 8 Helen Oakley
From Revolution to Migration. A Study of Contemporary Cuban and Cuban American Crime Fiction. *Forthcoming.*
ISBN 978-3-03911-021-6

Vol. 9 Thea Pitman
Mexican Travel Writing. 209 pages. 2008.
ISBN 978-3-03911-020-9

Vol. 10 Francisco J. Borge
A New World for a New Nation. The Promotion of America in Early Modern England. 240 pages. 2007.
ISBN 978-3-03911-070-4

Vol. 11 Helena Buffery, Stuart Davis and Kirsty Hooper (eds)
Reading Iberia. Theory/History/Identity. 229 pages. 2007.
ISBN 978-3-03911-109-1

Vol. 12 Sheldon Penn
Writing and the Esoteric. Identity, Commitment and Aesthetics in Mexican Literature since the Revolution. *Forthcoming.*
ISBN 978-3-03911-101-5

Vol. 13 Angela Romero-Astvaldsson
La obra narrativa de David Viñas. La nueva inflexión de *Prontuario* y *Claudia Conversa*. 300 pages. 2007.
ISBN 978-3-03911-100-8

Vol. 14 Aaron Kahn
The Ambivalence of Imperial Discourse. Cervantes's *La Numancia* within the 'Lost Generation' of Spanish Drama (1570–90).
243 pages. 2008.
ISBN 978-3-03911-098-8

Vol. 15 Turid Hagene
Negotiating Love in Post-Revolutionary Nicaragua. The role of love in the reproduction of gender asymmetry. 341 pages. 2008.
ISBN 978-3-03911-011-7

Vol. 16 Yolanda Rodríguez Pérez
The Dutch Revolt through Spanish Eyes. Self and Other in historical and literary texts of Golden Age Spain (c. 1548–1673). 346 pages. 2008.
ISBN 978-3-03911-136-7

Vol. 17 Stanley Black (ed.)
Juan Goytisolo: Territories of Life and Writing. 202 pages. 2007.
ISBN 978-3-03911-324-8

Vol. 18 María T. Sánchez
The Problems of Literary Translation. A Study of the Theory and Practice of Translation from English into Spanish. 269 pages. 2009.
ISBN 978-3-03911-326-2

Vol. 19 Matías Bruera
Meditations on Flavour. *Forthcoming.*
ISBN 978-3-03911-345-3

Vol. 20 Ana Cruz García
Re(de-)generando identidades. Locura, feminidad y liberalización en Elena Garro, Susana Pagano, Ana Castillo y María Amparo Escandón. 259 pages. 2009.
ISBN 978-3-03911-524-2

Vol. 21 Idoya Puig (ed.)
Tradition and Modernity. Cervantes's Presence in Spanish Contemporary Literature. 221 pages. 2009.
ISBN 978-3-03911-526-6

Vol. 22 Charlotte Lange
Modos de parodia. Guillermo Cabrera Infante, Reinaldo Arenas, Jorge Ibargüengoitia y José Agustín. 252 pages. 2008.
ISBN 978-3-03911-554-9

Vol. 23 Claudio Canaparo
Geo-epistemology. Latin America and the Location of Knowledge. 284 pages. 2009.
ISBN 978-3-03911-573-0

Vol. 24 Jesús López-Peláez Casellas
"Honourable Murderers". El concepto del honor en *Othello* de Shakespeare y en los "dramas de honor" de Calderón. 321 pages. 2009.
ISBN 978-3-03911-825-0

Vol. 25 Marian Womack and Jennifer Wood (eds)
Beyond the Back Room. New Perspectives on Carmen Martin Gaite. *Forthcoming*.
ISBN 978-3-03911-827-4

Vol. 26 Manuela Palacios and Laura Lojo (eds)
Writing Bonds. Irish and Galician Contemporary Women Poets. 232 pages. 2009.
ISBN 978-3-03911-834-2

Vol. 27 Myriam Osorio
Agencia femenina, agencia narrativa. Una lectura feminista de la obra en prosa de Albalucía Ángel. 180 pages. 2010.
ISBN 978-3-03911-893-3

Vol. 28 Aino Linda Rinhaug
Fernando Pessoa: A Ludicrous Self. *Forthcoming.*
ISBN 978-3-03911-909-7

Vol. 29 Soledad Pérez-Abadín Barro
Cortázar y Che Guevara: Lectura de *Reunión.* 182 pages. 2010.
ISBN 978-3-03911-919-6

Vol. 30 Gonzalo Pasamar Alzuria
Construsting National Narratives: Historians and Historiography in Spain. *Forthcoming.*
ISBN 978-3-03911-920-2

Vol. 31 Victoria Carpenter (ed.)
(Re)Collecting the Past: History and Collective Memory in Latin American Narrative. 315 pages. 2010.
ISBN 978-3-03911-928-8

Vol. 32 Geneviève Fabry, Ilse Logie and Pablo Decock (eds)
Imaginarios apocalípticos en la literatura hispanoamericana. *Forthcoming.*
ISBN 978-3-03911-937-0

www.ingramcontent.com/pod-product-compliance
Ingram Content Group UK Ltd.
Pitfield, Milton Keynes, MK11 3LW, UK
UKHW021852100926
8556IPUK00020B/1182

9 783039 115730